Convexity in discrete structures

Ramanujan Mathematical Society Lectures Notes Series

Vol. 1: Number theory

Vol. 2: The Riemann zeta function and related themes

Vol. 3: Formal language aspects of natural computing

Vol. 4: Commutative algebra and combinatorics

Vol. 5: Convexity in discrete structures

Vol. 6: Number theory and discrete geometry

Vol. 7: Discrete mathematics

Vol. 8: Lectures on operator theory

Vol. 9: Essays on geometric group theory

Vol. 10: Techmüller theory and moduli problems

Ramanujan Mathematical Society

Lecture Notes Series

Volume 5

Convexity in discrete structures

Proceedings of the international instructional workshop on convexity in discrete structures, Thiruvananthapuram, India, 2006; and the international workshop on metric and convex graph theory, Barcelona, Spain, 2006

Volume editors

Manoj Changat
Sandi Klavzar
Henry Martyn Mulder
A. Vijayakumar

International Press
www.intlpress.com

Ramanujan Mathematical Society
Lecture Notes Series, Volume 5
Convexity in discrete structures

Volume Editors:
Manoj Changat
Sandi Klavzar
Henry Martyn Mulder
A. Vijayakumar

Published in 2010 by International Press, Somerville, Massachusetts, U.S.A. under license from the Ramanujan Mathematical Society.

This work is also published in the Republic of India, exclusively by the Ramanujan Mathematical Society.

ISBN 978-1-57146-190-2

Typeset using the LaTeX system.
Printed in the United States of America.

Contents

Foreword

This special volume of the Ramanujan Mathematical Society Lecture Note Series is the joint proceedings of the International Instructional Workshop on Convexity in Discrete Structures held at Thiruvananthapuram, Kerala, India (March 22–April 2, 2006) and the International Workshop on Metric and Convex Graph Theory held at Barcelona, Spain (June 12–16, 2006). The volume contains 12 selected refereed articles (surveys and contributed papers) on which talks at the conference were based. They are all devoted to abstract and discrete convexity and to discrete and graphic metric spaces.

The paper by B. D. Acharya reviews the set systems which can be realized as families of neighborhoods and iterated neighborhoods of graphs and their relationships with abstract interval structures. Four papers are devoted to the investigation of various transit functions on graphs, posets, and lattices, another structure closely related to intervals, convexity, and betweenness. The paper by H. M. Mulder presents the general theory of transit functions and formulate the open questions in this area. A. Mathews and J. Mathews investigate the transit functions on posets and lattices, while M. Changat and J. Mathews characterize the graphs for which the induced path transit function is monotone. Finally, Prasanth G. N. studies a transit function closely related with the geodesic transit function on graphs. P. Duchet investigates the convexity parameters of segment spaces and presents a sharp upper bound of Radon numbers of these spaces in terms of their Helly numbers. The paper by Kannan B. and M. Changat surveys the results about the hull number and the rank of induced path and triangle-path convexities of graphs. The paper by I. M. Pelayo overview the results on extremal vertices and Krein-Milman property of various graph-convexities. K. Reji Kumar investigates the convexity property of minimal edge dominating functions of graphs. C. Hernando, M. Mora, P. Slater, and D. R. Wood present a relationship between the fault-tolerant metric dimension and the classical metric dimension of graphs and an investigation of the first concept for trees. S. Klavžar presents a survey of recent results on the structure and the constructions of cubical partial cubes and formulates some remaining open questions. Finally, M. Changat and V. Chepoi give an overview of classical and recent results in the theory of finite metric spaces, their embedding and related graph classes.

In conclusion, the proceedings present a valuable collection of original and survey papers in the growing area of convexity studies and metric graph theory. Therefore the collection will be a welcome source for researchers in these areas.

Marseille, France
September 2007

Victor Chepoi

Acknowledgement

The International Instructional Workshop on Convexity in Discrete Structures (IIWCDS) was organized by the Department of Futures studies, University of Kerala, Thiruvananthapuram from March 22 to April 2, 2006 at Kovalam, Thiruvananthapuram, India. The workshop was sponsored by the Department of Science and Technology (DST), Govt. of India and co-sponsored by University of Kerala with its University Grants Commission's (UGC) unassigned grant and Kerala State Council for Science, Technology and Environment (KSCSTE), Govt. of Kerala. We express our sincere gratitude to DST, UGC and KSCSTE. The authorities of the University of Kerala and the staff and research scholars of the Department of Futures Studies have helped wholeheartedly for the success of the workshop. Many thanks to all of them for making this a fruitful academic reality.

The International Workshop on Metric and Convex Graph Theory (IWMCGT) was organized by Ignacio M. Pelayo and Carlos Seara and hosted by the research group in Computational Geometry and Combinatorial Geometry of the Polytechnic University of Catalonia from June 12th to 16th, 2006 in Barcelona, Spain. The workshop was sponsored by research group in Computational Geometry and Combinatorial Geometry, Applied Mathematical Department II, and by the Polytechnic University of Catalonia. We express our sincere gratitude to the UPC research group in Computational Geometry and Combinatorial Geometry for hosting and sponsoring the workshop.

Manoj Changat
Coordinator, IIWCDS

Ignacio M. Pelayo & Carlos Seara
Organizers, IWMCGT

Preface

Convexity is basically a branch of geometry. It has innumerable connections with the other areas of mathematics such as analysis, algebra, topology, and combinatorial optimization. One of the most attractive aspects of convexity theory is the large number of easily stated and intuitively appealing unsolved problems that it still contains. The significance of convexity theory is that it deals analytically with global properties of objects and enables one to deal with extremal problems. There is a vast amount of literature on convexity theory from different perspectives. The axiomatic approach to convexity developed in the works of Levi, Jamison, Sierksma, Van de Vel and others, caught the attention of discrete mathematicians, especially graph theorists and resulted in vast literature of convexities in discrete structures like graphs, ordered sets, lattices etc. The myriad applications of convexity in discrete optimization, computational geometry etc. makes it a very interesting and important theme for discussion. The two workshops, one in India and the other in Spain, were held to discuss the problems on convexity mainly in graphs and other discrete structures.

This volume is the joint proceedings of the International Instructional Workshop on Convexity in Discrete Structures held at Thiruvananthapuram, Kerala, India during March 22–April 2, 2006 and the International Workshop on Metric and Convex Graph Theory held at Barcelona, Spain during June 12–16, 2006. The articles in this volume consist of the summary of lectures delivered and discussed in both workshops. There are 12 refereed articles consisting of surveys and contributed papers. Article 5 and article 7 are contributions from the Barcelona workshop and the rest are from the Thiruvananthapuram workshop.

With deep sense of gratitude, we register our indebtedness to the Ramanujan Mathematical Society (RMS), India for bringing out this joint proceedings. We are also extremely thankful to all the referees who devoted their valuable time for critically analyzing the articles and all the authors who contributed articles to this volume.

March 2007

Manoj Changat
Sandi Klavžar
Henry Martyn Mulder
A. Vijayakumar
(Editors)

Joint Proc. Int. Wrk.shps. Conv. in Discr. Struct. and Metric & Conv. Graph Th. No. 5, 2008, pp. 1–10.

Multiset Systems: A Creative Review

B. D. Acharya

Department of Science & Technology, Government of India, 'Technology Bhavan', New Mehrauli Road, New Delhi 110 016, India
e-mail: bdacharya@yahoo.com

Abstract. A family of 'submultisets' of a given nonempty set U is the object of study in this paper. As special cases, graphs whose open neighborhood families form conformal multiset systems have been characterized, and an update on an old problem of C. Berge (1974) requiring determination of graphs whose closed neighborhood families form balanced hypergraphs is given; the former solution is seen to have interesting connections with the notion of convexity in discrete structures as also with the problem of neighborhood reconstruction of the bond structure of chemical molecules in the spirit of the seminal works of P. Catlin (1978). Several theoretically interesting problems related to the vertex neighborhood families (VNFs) and multiset systems in general have been suggested in the sequel.

Keywords. Multiset, submultiset, vertex neighborhood family, interval structure, convexity, hypergraph, general system.

2000 Subject Classification: 05C65, 05C99, 52A01

1. Introduction

For standard terminology and notation in graph theory and hypergraph theory, we refer the reader to the classic text-books of F. Harary (1969) and C. Berge (1973) respectively, but the term 'clique' will be meant in the sense used in the latter reference.

Following J. E. Graver & M. E. Watkins (1977), if $\mathcal{N}$ denotes the set of nonnegative integers, given a nonempty set U, a function $s \in \mathcal{N}^U$ from U into $\mathcal{N}$, written $s : U \to \mathcal{N}$, is called a *selection from U*. If $x \in U$, we shall regard $s(x)$ as the number of times x is 'selected by s' and denote by $|s|$ the sum

$$\sum_{x \in U} s(x), \tag{1}$$

which is well-defined as a nonnegative integer; for instance, if $s(x) \leq 1$ for every $x \in U$ then $|s| = \sum_{x \in U} s(x) =: |A|$, the cardinality of the set $A = \{a \in U : s(a) = 1\}$. In general, we call $|s|$ the *weight* of s. If $|s| = m$, then we call s an *m-selection.* If $S_m(U)$ denotes the set of all m-selections from U then we have

$$\mathcal{N}^U := \bigcup_{m=0}^{\infty} S_m(U) =: S(U). \tag{2}$$

Note that if one regards a selection as a multiset

$$\langle y^{s(y)} : y \in U \rangle \tag{3}$$

where $y^{s(y)}$ denotes the string consisting of $s(y)$ y's for each of which $s(y) \neq 0$ then the standard *power set* 2^U of U can be regarded as a subset of $S(U)$.

The above notions suggest a way to generalize the idea of a 'system' from J. E. Graver & M. E. Watkins (1977) as follows: A *multiset system* (earlier called a *general system*; *cf.*, B. D. Acharya, 1983) H is an ordered triple (V, f, E) where V and E are disjoint sets and $f : E \to S(V)$ is a function specifying a family $\mathcal{F}_f$ of selections from V. Then, $H = (V, f, E)$ is a *system*, in the sense used by J. E. Graver & M. E. Watkins (1977), whenever every member of $f[E] := \langle f(e) : e \in E \rangle$ is a subset of V, that is, whenever $f[E]$ is a *family* of *subsets* of V.

The elements of E in a *multiset system* $H = (V, f, E)$ are called *multiblocks* of H and the elements of V are called the *vertices* of H. The *size* of a multiblock e of H is the weight $|f(e)|$ of $f(e)$ as defined in (1) above. In this way, almost all the terms, notations and definitions for 'set systems' given in J. E. Graver & M. E. Watkins (1977) can be easily understood and used freely unless a fresh outlook is warranted. To simplify our treatment here onwards, by a 'multiset system' we shall formally mean an ordered pair $H = (V, \mathcal{W})$, $\mathcal{W} = (N_i)_{i \in M}$ where M is an indexing set and N_i is a selection from V for each $i \in M$; in case N_i is a finite submultiset of V for each $i \in M$ then we say that H is *block finite* and, on the other hand, if the string $y^{s(y)}$ is finite for every $y \in N_i$, $\forall i \in M$ then we shall say that H is *locally finite*.

Observation 1.1. *Every block finite multiset system is locally finite, but not conversely.*

A fairly familiar example of a multiset system is the family $\mathcal{W}_G = \langle N(u) : u \in V(G) \rangle$ of the *vertex neighborhoods* $N(u) := \langle v \in V(G) : uv \in E(G) \rangle$ of the vertices of a *pseudograph* $G = (V, E)$ (*i.e.*, a *hypergraph* of *rank* 2). The family $\mathcal{W}_G$ is called the *vertex neighborhood family* (or simply, 'VNF' as in P. Catlin, 1978).

We shall call a multiset system $H = (V, \mathcal{W})$ *countable* whenever its vertex set V and its *selection family* $\mathcal{W}$ are both countable. Henceforth in this paper, we shall use the term *multiset system* to mean only that which is both countable and locally finite so that we can take $V = \{v_1, v_2, \dots\}$ and the selection family $\mathcal{W} = (N_i)_{i \in M}$ where M is an indexing set. Such a multiset system may be represented by its *incidence matrix* $I(H) = (a_{ij})$ whose i-th row corresponds to the vertex v_i of H, j-th column corresponds to the selection N_j and a_{ij} is the number of occurrences of v_i in N_j. We note here that in general $I(H)$ could be an infinite matrix whose entries are nonnegative integers. One may also represent H by its *incidence bigraph* $F = (V, M)$ which is a bipartite multigraph with bipartition $\{V, M\}$ where $v \in V$ is joined to $j \in M$ by as many edges as the number of occurrences of v in N_j. Note that if every selection $N_j \in \mathcal{W}$ is an ordinary subset of V then $H = (V, \mathcal{W})$ is a hypergraph, possibly infinite or having multiple edges, $I(H)$ is its usual $\{0, 1\}$-incidence matrix and $F = (V, M)$ is a simple bipartite graph (in general, a *simple* graph is a simple hypergraph of rank 2 having no *loops*).

2. Neighborhood systems

We shall call a multiset system $N = (V, \mathcal{W}),\ \mathcal{W} = (N_i)_{i \in M}$ a *neighborhood system* whenever there exists a bijection $\gamma : V \to M$ satisfying the *symmetry condition* **(S.C.)**: Every vertex v occurs as many times in $N_{\gamma(u)}$ as u occurs in $N_{\gamma(v)}$. Then, $\mathcal{W}$ is a VNF and we call γ a *neighborhood assignment* of N.

As noted in the foregoing section, given any pseudograph $G = (V, E)$, the family $N_G = \langle N(u) : u \in V(G)\rangle$ of the vertex neighborhoods $N(u) := \langle v \in V(G) : uv \in E(G)\rangle$ of the vertices of G is a neighborhood family, where for each $u \in V,\ N(u) =: N_{\gamma(u)}$ is the selection of all vertices v adjacent to u, with v appearing as many times as there are edges joining u to v in G. Then, $N_G = (V, \mathcal{W}_G),\ \mathcal{W}_G = (N(u))_{u \in V}$ is clearly a neighborhood system where $u \in N(u)$ if and only if there is the loop $\{u\}$ at u in G.

If V and E are nonempty sets and $f : E \to S(V)$, the function $f^* : V \to S(E)$ satisfying the *duality condition* **(D.C.)**: $e \in E$ occurs in $f^*(x)$ as many times as x occurs in $f(e)$, is called the *transpose* of f. Clearly, $f^{**} := (f^*)^* = f$. For any multiset system $H = (V, f, E)$, the multiset system $H^* = (E, f^*, V)$ is called the *dual* of H. Since $f^{**} = f$ we see that $H^{**} := (H^*)^* = H$. It is then easily seen that $I(H^*) = (I(H))^T$, the usual *transpose* of the matrix $I(H)$.

Given a function $f : X \to Y$ and a (locally finite) selection A from X, we denote by $f[A]$ the (locally finite) selection $\langle f(x) : x \in A\rangle$ from Y in which $f(x)$ occurs as many times as x occurs in A. Two multiset systems $H = (V, f, E)$ and $K = (W, g, F)$ are *isomorphic*, written $H \cong K$, if there exist bijections $\alpha : E \to F$ and $\beta : V \to W$ such that

$$\beta[f(e)] = g(\alpha(e)),\ \forall e \in E. \tag{4}$$

We shall call the ordered pair (α, β) an *isomorphism* from H onto K and a multiset system *self-dual* if it is isomorphic to its dual.

Theorem 2.1 (B. D. Acharya, 1983). *A countable multiset system is a neighborhood system if and only if it is self-dual.*

Proof. Let $N = (V, W),\ W = (N_i)_{i \in M}$, be any neighborhood system and $\gamma : V \to M$ be a neighborhood assignment of N. Since γ is a bijection such that **S.C.** is satisfied, it follows that the incidence matrix $I(N)$ is symmetric, and hence $I(N) = (I(N))^T = I(N^*)$ which means that $N \cong N^*$.

Conversely, if $N \cong N^*$ then the incidence matrices $I(N)$ and $I(N^*)$ of N and N^* respectively are 'equivalent' in the sense that one can be obtained from the other by some sequence of permutations of rows and/or columns. But since $I(N^*) = (I(N))^T$ we see that $I(N)$ and $(I(N))^T$ are equivalent so that there exists an automorphism π of N such that $I(N') = (I(N'))^T$ where N' is the automorphic image of N under π, written $\pi(N) = N'$ (note that $\pi = (\alpha, \beta)$ is an automorphism of N). Hence, $I(N') = (a_{ij})$ is a symmetric nonnegative matrix. Now, define a pseudograph $G = (V, E)$ where $V = \{v_1, v_2, \ldots\}$ and E by the rule: v_i and v_j are joined by a_{ij} edges, for any two vertices $v_i, v_j \in V$. Then, $I(N')$ is the *adjacency matrix* of the pseudograph G (henceforth adjacency matrix of G will be written as $A(G)$ in consonance with its standard notation). Now, it is not hard to see that $\mathcal{W}_G = \mathcal{W}'$, the family of multiblocks of N'. But since $N' = \pi(N)$, it follows that $\mathcal{W}' = \mathcal{W}$ so that $\mathcal{W}_G = \mathcal{W}$, and hence $N = N_G$. Thus, N is a neighborhood system. □

With any nonnegative integer matrix $A = (a_{ij})$ one can associate a bipartite multigraph $TB(A)$, called the *term bigraph* of A (*e.g., see* C. R. Cook 1970; C. R. Cook *et al.* 1974; S. T. Hedetniemi 1970; B. Zelinka, 1975 for this definition where A is a $\{0, 1\}$-matrix), as follows: The vertex set of $TB(A)$ is comprised of the set $\mathcal{R}_A = \{r_i\}$ of the rows of A and the set $\mathcal{C}_A = \{c_i\}$ of the columns of A. The vertex r_i and the vertex c_j are joined by a_{ij} distinct single edges, for all indices i and j for which (i, j) entry in A is not zero. Lastly, no two vertices within any of the sets $\mathcal{R}_A$ and $\mathcal{C}_A$ are defined to be adjacent.

The *tensor product* (or *Kronecker product* or *conjunction*; *e.g., see* E. Sampathkumar, 1981 or F. Harary, 1969) of two pseudographs G_1 and G_2, written $G_1 \wedge G_2$, is that graph whose adjacency matrix is the standard tensor product $A(G_1) \otimes A(G_2)$ of the adjacency matrices $A(G_1)$ and $A(G_2)$. In graph-theoretical terms, the vertex set of $G_1 \wedge G_2$ is the Cartesian product $V_1 \times V_2$ where any two vertices (u_i, v_j) and (u_s, v_t) are joined by as many single edges as the product of the number of edges joining u_i and u_s in G_1 and the number of edges joining v_j and v_t in G_2, every loop in either of the pesudographs being counted exactly once.

The automorphism π of N mentioned in the proof of Theorem 2.1, as a matter of fact, induces an order two automorphism β_π of $TB(I(N))$ that interchanges the set $\mathcal{R}_A$ of 'row-vertices' and the set $\mathcal{C}_A$ of 'column-vertices' respectively. We note here that not all automorphisms of N induce such an automorphism of $TB(I(N))$. But, every automorphism π of N whose image N' satisfies $I(N') = (I(N'))^T$, if it exists, determines a pseudograph G_π such that

$$A(G_\pi) = I(N') \tag{5}$$

and also

$$TB(I(N)) = G_\pi \wedge K_2 \tag{6}$$

and each such automorphism of N induces an order two automorphism of $TB(I(N))$ which interchanges $\mathcal{R}_{I(N)}$ and $\mathcal{C}_{I(N)}$. However, as observed by P. Catlin (1978), there might exist two different automorphisms π and π' of N with the above properties such that G_π and $G_{\pi'}$ are not isomorphic; in fact, he gave algebraic necessary and sufficient conditions for a hypergraph H to be the neighborhood system of some pseudograph as also an algebraic formula for the number of nonisomorphic pseudographs each of whose neighborhood systems is represented by H (P. Catlin, 1978a, 1978b).

The above analysis remains to be extended for multiset systems which might, for instance, represent the neighborhood systems of pseudographs. We note here, in passing, that nonisomorphism between two pseudographs may arise even due to the difference in multiplicities of the edges. Hence, Catlin's '*neighborhood reconstruction problem*' may now be stated specifically for multigraphs as follows: Which multigraphs are uniquely reconstructible (or, 'representable'), up to isomorphism, from (by) their VNFs? As mentioned before, this problem has already been solved completely by

P. Catlin (1978a, 1978b) for pseudographs, but it remains open for multigraphs which are not pseudographs. Solution of this specific problem has far-reaching implications in polymer chemistry (*e.g., see* S. C. Basak *et al.* 2007).

3. Neighborhood hypergraphs

A *neighborhood hypergraph* $N = (V, \mathcal{W})$ is one for which there exists a pseudograph $G = (V, E)$, with at most one loop at any of its vertices, such that $N \cong N_G = (V, \mathcal{W}_G)$. We present here some combinatorial characterizations of such hypergraphs. Interestingly, as we shall see shortly, they are closely related to the notion of convexity.

Theorem 3.1 (B. D. Acharya, 1983). *A hypergraph is a neighborhood hypergraph if and only if it is self-dual.*

Since by a theorem of C. Berge (1973; p. 399), a hypergraph is a Helly hypergraph if and only if its dual is conformal, we get

Theorem 3.2 (B. D. Acharya, 1983). *A neighborhood hypergraph is conformal if and only if it satisfies the Helly property.*

Let V be a finite set. Then, a mapping $\psi : V \times V \to 2^V$ is called an *interval structure* on V if it satisfies

$$x, y \in \psi(u, v) \Leftrightarrow \psi(x, y) \subseteq \psi(u, v), \ \forall x, y, u, v \in V \tag{7}$$

and

$$\psi(u, v, w) := \psi(u, v) \cap \psi(v, w) \cap \psi(w, u) \neq \emptyset, \ \forall u, v, w \in V. \tag{8}$$

If ψ is an interval structure on V, each set $\psi(u, v)$ is called an *interval* in V (H. M. Mulder 1980). A subset X of V is then called ψ*-convex* if, for any two elements $u, v \in X$, the ψ-interval $\psi(u, v)$ is contained in X (for graphs, this idea was introduced by B. D. Acharya, 1972; *also, see* B. D. Acharya *et al.* 1973 and R. E. Jamison 1986).

Next, given any hypergraph $H = (X, \mathcal{E})$ and any two vertices $x, y \in X$, let

$$\psi_{\mathcal{E}}(x, y) := \cap\{B \in \mathcal{E} : x, y \in B\} \tag{9}$$

Then, the following reformulation of a result originally due to P. C. Gilmore (*cf.*: C. Berge, 1973; p. 396) is found in H. M. Mulder (1980; p. 98).

Theorem 3.3. *A hypergraph $H = (X, \mathcal{E})$ has the Helly property if and only if $\psi_{\mathcal{E}}$ is an interval structure on X.*

The following result is then an easy deduction from Theorems 3.2, 3.3.

Theorem 3.4. *A neighborhood hypergraph $N = (V, \mathcal{W})$ is conformal if and only if $\psi_{\mathcal{W}}$ is an interval structure on V.*

Corollary 3.5. *Let $N = (V, \mathcal{W})$ be a conformal neighborhood hypergraph. Then, any family $\mathcal{E}$ of nonempty $\psi_{\mathcal{W}}$-convex subsets of V has the Helly property.*

Proof. Since N is conformal, $\mathcal{W}$ satisfies the Helly property and hence, by Theorem 3.4, $\mathcal{W}$ is an interval structure on V. Hence the result follows from a theorem of Mulder (*cf.*: 1980; p. 98, Corollary 4.1.3). □

4. Distance-bound neighborhood systems of graphs

Given any finite connected multigraph $G = (V, E)$, any positive integer r and a vertex u in G, the *closed r-neighborhood* of u, denoted $C_r(u)$, is defined to be the multiset of all vertices in G which are at distance at most r from u and the set $N_r(u) = C_r(u) - \{u\}$ is called the *open r-neighborhood* of u in G. Hence, we let $\mathcal{C}_r(G) = \langle C_r(u) : u \in V(G)\rangle$ and $\mathcal{N}_r(G) = \langle N_r(u) : u \in V(G)\rangle$. Then, the multiset systems defined by these VNFs of G on $V(G)$ are called the *closed r-neighborhood* and the *open r-neighborhood multiset systems*, respectively. Given any property of multiset systems, say $\mathcal{P}$, one may then be interested to characterize the multigraphs G for which $\mathcal{C}_r(G)$ or $\mathcal{N}_r(G)$ has the property $\mathcal{P}$ (*e.g., see* C. Berge, 1973; p. 278, in the case of hypergraphs). These *distance-bound* neighborhood systems too are being used to devise better mathematical descriptors of structural characteristics of chemical molecules (*e.g., see* S. C. Basak *et al.* 2007).

For making this article self-contained, we shall give outlines of the proofs of the main results appearing in this section.

By a *section* of a multiset system $H = (X, \mathcal{E})$ we mean a multiset system $H' = (X, \mathcal{E}')$ such that for every $E' \in \mathcal{E}'$ there exists a selection $E \in \mathcal{E}$ with $E' \subseteq E$. Further, $H^u = (X, \mathcal{W}^u)$, $\mathcal{W}^u = \{N_i^u\}_{i \in M}$, N_i^u being the set of distinct elements that occur in N_i, is called the *underlying hypergraph* of H.

In this section, we shall consider a specific instance of the general problem of characterizing simple connected graphs whose open r-neighborhood multiset systems are conformal. Some preliminary observations toward this end are imperative and are stated in the next proposition (*cf.*: B. D. Acharya 1985), where 'd' denotes the *distance metric* on $V(G)$.

Theorem 4.1. *Let $G = (V, E)$ be any graph and r be any positive integer. Then, the following assertions hold:*

(i) $u \in N_r(v) \Leftrightarrow v \in N_r(u), \forall u, v \in V(G)$
(ii) *$\mathcal{N}_r(G)$ is self-dual*
(iii) *$\mathcal{N}_r(G)$ is conformal if and only if $\mathcal{N}_r(G)$ is a multiset system whose underlying hypergraph satisfies the Helly property*
(iv) *For $r \geq 2$, $N_{r-1}(u) \subseteq N_r(u)$ for each $u \in V(G)$*
(v) *For $r \geq 2$, $\mathcal{N}_{r-1}(G)$ is a section of $\mathcal{N}_r(G)$.*

Lemma 4.2. *For any graph G and any positive integer r, if $\mathcal{N}_r(G)$ is conformal then G has no cycle of length $\leq 3r$.*

Proof. Suppose that $\mathcal{N}_r(G)$ is conformal and that G does have a cycle Z of length at most $3r$. Then, we may choose any three vertices u_1, u_2, u_3 on Z so that $d(u_1, u_2) \leq r, d(u_2, u_3) \leq r$ and $d(u_3, u_1) \leq r$. This implies $u_1, u_2 \in N_r(u_3)$, $u_2, u_3 \in N_r(u_1)$ and $u_1, u_3 \in N_r(u_2)$. Hence, $\{u_1, u_2, u_3\}$ would be a clique in the 2-section of $\mathcal{N}_r(G)$ such that $\{N_r(u_1), N_r(u_2), N_r(u_3)\}$ has pairwise intersection property. Let Q be a clique of maximum size in the 2-section of $\mathcal{N}_r(G)$ such that the family $\{N_r(u) : u \in Q\}$ has the pairwise intersection property. As seen above, $|Q| \geq 3$. Since $\mathcal{N}_r(G)$ is conformal we see from Theorem 4.1(iii), that $\mathcal{N}_r$ must satisfy the Helly property and hence as $\{N_r(u) : u \in Q\}$ has pairwise intersection property we must have a vertex $x \in \bigcap_{u \in Q} N_r(u)$. Then, by Theorem 4.1(i), $u \in N_r(x)$ for every $u \in Q$ so that $Q \subseteq N_r(x)$. Since $x \notin N_r(x)$ we have $x \notin Q$. Let $Q_x = \{x\} \cup Q$. Since $|Q| \geq 3$, for any given $u \in Q$ there exists $u' \in Q - \{u\}$ such that $x, u \in N_r(u')$ so that xu is an edge in the 2-section of $\mathcal{N}_r(G)$. Thus, x is adjacent to every vertex of Q in the 2-section of $\mathcal{N}_r(G)$ and hence Q_x is a clique of $(\mathcal{N}_r(G))_{(2)}$ such that $\{N_r(y) : y \in Q_x\}$ has the pairwise intersection property. But, this contradicts the maximality of Q as $|Q_x| = |Q| + 1$. Hence, the assertion follows. □

Lemma 4.3. *For any graph G, if $\mathcal{N}_r(G)$ is conformal for any integer $r \geq 2$ then no two edges of G are adjacent.*

Proof. Suppose that uv and vw are two incident edges in G and that $\mathcal{N}_r(G)$ is conformal for some positive integer $r \geq 2$. Then, it is easy to see that $v, w \in N_r(u)$, $u, w \in N_r(v)$ and $u, v \in N_r(w)$, and hence $\{u, v, w\}$ is a clique in the 2-section of $\mathcal{N}_r(G)$. Since $N_r(G)$ is conformal it follows that there must exist $x \in V(G)$ such that $\{u, v, w\} \subseteq N_r(x)$. This implies that the subgraph $\langle N_r(x) \rangle$ induced by $N_r(x)$ in G has a cycle of length $\leq 2r + 1 < 3r$ as $r \geq 2$. But, this is contrary to Lemma 4.2. □

Theorem 4.4 (B. D. Acharya 1985). *For any finite graph G, $\mathcal{N}(G)$ is conformal if and only if G has neither C_3 (the 'triangle') nor C_6 (the 'hexagon') as an induced subgraph.*

Proof. Necessity: Let $\mathcal{N}(G)$ be conformal. Then, by Lemma 4.2, G has no induced subgraph isomorphic to C_3. Hence, suppose that G has $C_6 = (u_1, u_2, u_3, u_4, u_5, u_6, u_1)$ as an induced subgraph. Then, $u_2, u_6 \in N(u_1)$, $u_1, u_3 \in N(u_2)$, $u_2, u_4 \in N(u_3)$, $u_3, u_5 \in N(u_4)$, $u_4, u_6 \in N(u_5)$, $u_5, u_1 \in N(u_6)$. Thus, $\{u_1, u_3, u_5\}$ is a clique in the 2-section of $\mathcal{N}(G)$ such that $\{N(u_1), N(u_3), N(u_5)\}$ has pairwise intersection property in G. Hence, we let Q be a clique of maximum cardinality in the 2-section of $\mathcal{N}(G)$ such that $\{N(u) : u \in Q\}$ has pairwise intersection property in G. Then, as in the proof of the foregoing result, using conformality of $\mathcal{N}(G)$ we are led to a contradiction to the maximality of Q.

Sufficiency: Suppose that G has no induced triangles or hexagons, but that $\mathcal{N}(G)$ is not conformal. Then, by applying a theorem of Gilmore (*cf.* C. Berge 1973; p. 396) to $\mathcal{N}(G)$ it follows that there must exist $u, v, w \in V(G)$ such that $\{N(u), N(v), N(w)\}$ has pairwise intersection property while the set $N(u, v, w)$ is not contained in the open neighborhood of any vertex of G. Let $x \in N(u) \cap N(v)$, $y \in N(v) \cap N(w)$ and let $z \in N(w) \cap N(u)$. Since G has no triangles, it is easy to see that the sets $\{u, v, w\}$ and $\{x, y, z\}$ must be independent sets in G (as also the open neighborhoods of vertices in them). Since G does not contain any hexagon either, we see that at least one of the pairs, *viz.* $\{u, y\}$, $\{v, z\}$ and $\{w, x\}$ must be an edge of G; say $uy \in E(G)$. Then, $\{x, y, z\} \subseteq N(u)$. Since $N(u, v, w) \not\subseteq N(u)$ there must exist $y' \in N(v) \cup N(w) - \{y\}$ such that $uy' \notin E(G)$. Now, if neither $\{w, x\}$ nor $\{v, z\}$ is an edge of G, then $\langle\{x, v, y', w, z, u\}\rangle$ is an induced hexagon, so assume $vz \in E(G)$. This yields, $\{x, y, z\} \subseteq N(v)$; since $N(u, v, w) \nsubseteq N(v)$, there must exist $z' \in N(u) \cup N(w) - \{z\}$ such that $vz' \notin E(G)$. Then, if $wx \notin E(G)$ we obtain the induced hexagon $\langle\{x, v, y', w, z', u\}\rangle$ in G. Thus, $wx \in E(G)$ and hence $\{x, y, z\} \subseteq N(w)$. Since $x \in N(u) \cap N(v)$, $y \in N(v) \cap N(w)$

and $z \in N(w) \cap N(u)$ are arbitrary by choice, we see that $N(u, v, w)$ is contained in each of $N(u)$, $N(v)$ and $N(w)$, a contradiction. Thus, $\mathcal{N}(G)$ must be conformal. □

Using the foregoing four results, the following characterization has been obtained.

Theorem 4.5 (B. D. Acharya 1985). *For any finite connected graph G and for any positive integer r, $\mathcal{N}_r(G)$ is conformal if and only if either (i) $r = 1$ and G has neither C_3 (the 'triangle') nor C_6 (the 'hexagon') as an induced subgraph, or (ii) $r \geq 2$ and no two edges of G are adjacent.*

5. Closed neighborhood systems of graphs

Recall that a hypergraph $H = (X, \mathcal{E})$ is *balanced* if every odd cycle $(x_1, E_1, x_2, E_2, \ldots, x_n, E_n, x_1)$ in H has an edge E_j containing at least three vertices x_i, $1 \leq i \leq n$ of the cycle. It is well known that a balanced hypergraph is conformal, but not conversely (*e.g., see* C. Berge 1973, 1986). Let e be a chord of an even cycle C in a graph G, dividing C into two paths P_1 and P_2. Then, e is called an *even chord* if both $\{e\} \cup P_1$ and $\{e\} \cup P_2$ are even cycles. Following the problem of characterizing a simple graph G whose closed neighborhood system constitutes a balanced hypergraph raised by C. Berge (1974), the following characterization was obtained.

Theorem 5.1 (R. Balakrishnan & K. Ranganathan 1981). *For any finite connected simple graph G, $\mathcal{C}(G)$ is balanced if and only if (i) G is triangulated and (ii) every cycle in G whose length is of the form $4m + 2$ has an even chord.*

The following result gives a sufficient condition for the closed neighborhood system to constitute a conformal hypergraph.

Theorem 5.2 (R. Balakrishnan & K. Ranganathan 1981). *For any finite connected simple graph G, having no chordless 5-cycles or 6-cycles, every partial hypergraph of $\mathcal{C}(G)$ is conformal; in particular, $\mathcal{C}(G)$ is conformal.*

The conditions of Theorem 5.2 are only sufficient but not necessary for $\mathcal{C}(G)$ to be conformal; for example, if $G = K_1 + (C_5 \cup C_6)$, $\mathcal{C}(G)$ is conformal (since $\mathcal{C}(G)$ has an edge containing all the vertices of G, every clique of $G^2 = (\mathcal{C}(G))_{(2)}$ is contained in this edge), even though G contains a 4-cycle, a chordless 5-cycle and a chordless 6-cycle. However, in the case of bipartite graphs, the conditions of Theorem 5.2 have been found to be necessary too.

Theorem 5.3 (R. Balakrishnan & K. Ranganathan 1981). *If G is a connected bipartite graph, then $\mathcal{C}(G)$ is conformal if and only if G has no chordless cycles of lengths 4 or 6.*

A little more general result than Theorem 5.3 is the following one.

Theorem 5.4 (R. Balakrishnan & K. Ranganathan 1981). *If G is a connected graph without quadrilaterals, then $\mathcal{C}(G)$ is conformal if and only if G has no chordless cycles of lengths 5 or 6.*

At length, R. Balakrishnan & K. Ranganathan (1981) have concluded with the following two important remarks that need to be kept in mind while attempting to completely determine graphs whose closed neighborhood systems form a conformal hypergraph:

(A) The condition that G be free from 4-cycles is an essential part of the hypothesis, for if $G = C_4$ then $\mathcal{C}(G)$ is not conformal.
(B) The conditions of the last three theorems, however, are not sufficient for $\mathcal{C}(G)$ to be balanced as, for instance, the cycles C_{2n}, $n \geq 4$, satisfy the conditions of these three theorems but, by the following result, $\mathcal{C}(C_{2n})$ is not balanced.

Theorem 5.5 (R. Balakrishnan & K. Ranganathan 1981). *If G is a cycle, then $\mathcal{C}(G)$ is balanced if and only if G is isomorphic to C_3, the triangle.*

The general problem of completely determining a graph whose closed neighborhood system forms a conformal (balanced) hypergraph is still open. Determining a graph whose open neighborhood system forms a balanced hypergraph is also open.

References

[1] B. D. Acharya, Distance-convex sets in graphs, Res. Rep. No. 4, IIT, Bombay (1972).
[2] B. D. Acharya, S. P. R. Hebbare and M. N. Vartak, d-convex sets in a graph, Graph Theory Newsletter (1973).
[3] B. D. Acharya, General systems and related problems, *Indian J. Pure & Appl. Math.*, **14** (1983) no. 9, 1083–1090.
[4] B. D. Acharya, Graphs whose r-neighborhoods form conformal hypergraphs, *Indian J. Pure & Appl. Math.*, **16** (1985) no. 5, 461–464.
[5] R. Balakrishnan and K. Ranganathan, Graphs whose neighborhood hypergraphs are balanced, In: Combinatorics and Applications (eds.) K. S. Vijayan and N. M. Singhi, Proc. Seminar on Combinatorics and Applications in honour of Prof. S. S. Shrikhande (14–17 Dec. 1982: Indian Statistical Institute, Kolkata), pp. 48–54. Indian Statistical Institute, Kolkata (1984).
[6] S. C. Basak, D. Mills and B. D. Gute, Predicting bioactivity and toxicity of chemicals from mathematical descriptors: A chemical-cum-biochemical approach, In: Advances in Quantum Chemistry: Chemical Graph Theory: Wherefrom, wherefor and whereto? (eds.) D. J. Klein and D. Brandas, 1–91. Elsevier-Academic Press (2007).
[7] C. Berge, Graphs and Hypergraphs, North-Holland Elsevier Publ. Comp., Amsterdam (1973)
[8] C. Berge, Hypergraphs, North-Holland Elsevier Publ. Comp., Amsterdam (1986).
[9] C. Berge, Hypergraph Seminar, Lecture Notes in Mathematics No. 411 (eds.) C. Berge and D. K. Ray-Chauduri, Springer-Verlag, Berlin (1974).
[10] P. A. Catlin, Vertex neighborhood families of graphs, Unpublished Manuscript (1978a).
[11] P. A. Catlin, Nonisomorphic graphs having the same vertex neighborhood family, Proc. IX S-E Conf. Combinatorics, Graph theory & Computing (Florida Atlantic University, Boca Raton, Fla., 1978b), 187–193. Congressus Numerantium XXI, Utilitas Mathematica, Winnipeg (1978).
[12] C. R. Cook, Graphs associated with {0, 1}-arrays, Themis Project Report No. 18 (Doctoral Dissertation), Univ. of Iowa (1970).
[13] C. R. Cook, B. D. Acharya and V. Mishra, Adjacency graphs, Proc. V S-E Conf. Combinatorics, Graph theory & Computing (Florida Atlantic University, Boca Raton, Fla., 1974), 317–331. Congressus Numerantium *X*, Utilitas Mathematica, Winnipeg (1974).

[14] J. E. Graver and M. E. Watkins, Combinatorics with emphasis on the theory of graphs, Springer-Verlag, GTM-54, New York (1977).
[15] F. Harary, Graph Theory, Addison-Wesley, Reading, Massachusetts (1969).
[16] S. T. Hedetniemi, Graphs of (0, 1)-matrices, In: Recent trends in Graph Theory, Springer-Verlag LNM-186, New York (1970).
[17] R. E. Jamison and M. Farber, Convexity in graphs and hypergraphs, *SIAM J. Algebraic Discrete Methods*, **7** (1986) 433–444.
[18] H. M. Mulder, The interval function of a graph, MC-Tract 132, Mathematisch Centrum, Amsterdam (1980).
[19] E. Sampathkumar, Tensor product of graphs, *J. Austr. Math. Soc.*, **20** (1981) 268–273.
[20] B. Zelinka, {0, 1}-matrices having isomorphic term bigraphs have equal ranks, *J. Indian Math. Soc.*, **XXXIX** (1975) 315–319.

Joint Proc. Int. Wrk.shps. Conv. in Discr. Struct. and Metric & Conv. Graph Th. No. 5, 2008, pp. 11–23.

Hull Numbers of Path Convexities on Graphs*

B. Kannan[1] and Manoj Changat[2]

[1]Department of Computer Applications, Cochin University of Science and Technology, Cochin 682 022, India
[2]Department of Futures Studies, University of Kerala, Thiruvananthapuram 695 034, India
e-mail: bkannan@cusat.ac.in; mchangat@gmail.com

Abstract. A feasible family of paths in a connected graph G such as the 'induced paths', 'triangle paths' and 'all paths' defines a convexity on G. We survey the behavior of such convexities on the hull number and rank. Along this, we present algorithms for computing the hull number, hull set and the convex hull for the induced path convexity using the decomposition algorithms by clique separators.

Keywords. Transit function, induced path, convex hull, hull number, clique separator.

Subject Classification: 05C99, 05C12, 05C38, 05C75, 52A01, 68R10

1. Introduction

The study of convexities in graphs is mainly focussed on convexities associated with interval functions, defined by paths. A prime example is the interval function $I : V \times V \to 2^V$ of a connected graph G, where $I(u, v)$ is the set of the vertices lying on shortest paths between u and v. Recently, starting from H. M. Mulder [22], the properties of the interval function such as betweenness, monotonicity and other problems are attempted under the term *transit functions*. The other important interval functions on graphs are the induced path, triangle-path and all-paths transit functions. See the following references for different aspects of such transit functions and associated convexities [3,4,6,9,10,18,20,21,23,25,29]. The behaviour of such convexities on the classical convexity invariants, such as the Carathéodory, Helly and Radon numbers in relation with related graph invariants has been surveyed in [7]. See also [10] for the classical convexity invariants of the induced path convexity, [11] for the geodesic convexity, and [3] for the triangle path convexity. In fact, Duchet in [11] underlines the importance of studying the geodesic convexity with graphs in regard to the classical convexity invariants Carathéodory (c), Helly (h) and partition numbers (p_k) with the following observation: For every convexity on a finite set V, with given c, h, and p_k, there is a finite connected graph G whose geodesic convexity has Carathéodory number (c), Helly number h, Partition number at least p_k. This motivates the study of convexity invariants of other convexities on graphs.

In this paper, we focus on the hull number and the rank. The hull number is an important graph parameter. There are many studies on the hull number for the geodesic

*Supported by the Department of Science and Technology, Govt. of India under the grant DST/INT/SLOV/P-03/05.

convexity. For example the relation between the hull number and geodetic number (another important graph invariant) for many classes of graphs has been discussed in [8], [13] and [24]. Similar study has been attempted for the induced path convexity, see for example, [15], [24]. In the case of the geodesic convexity, it is very difficult to estimate the hull number and the rank for arbitrary graphs since one can construct graphs with arbitrary large Carathéodory and Helly numbers. But the situation is very different for the induced path and triangle induced path convexities, as we have very good estimates for these classical invariants.

We attempt here to compute the hull number and the rank for the induced path and triangle induced path convexities in arbitrary graphs. In this process, we also design an algorithm to compute the minimum hull set and hull number of the graph, and the induced path convex hull of a subset of the vertex set. A decomposition of a connected graph using minimal clique separators is algorithmically established by Berry and Bordat [1]. Using the algorithm, one can yield a unique decomposition containing all the atoms of the graphs. Note that Tarjan's algorithm using clique separators also produces a decomposition of the graph which may not be unique [30] as Tarjan himself pointed out. Both the algorithms decompose a connected graph G into atoms and clique separators and the decomposition yields a decomposition tree whose leaves corresponds to atoms of the graph G. Using the algorithm in [1], we obtain a unique decomposition tree. This decomposition tree also produces the minimal set of leaves which satisfies our requirement for the minimal hull set.

In [1], Berry *et al.* have obtained an ordering of all atoms using what is known as moplex ordering. For completeness, we describe briefly the moplex ordering and the decomposition of a graph by minimal clique separators. We start with shortly recalling the various definitions.

We consider only finite connected, simple and loopless graphs $G = (V, E)$. By an (m, n)-graph, we mean a graph with m vertices and n edges.

2. Preliminaries

A *transit function* on G is a function $R : V \times V \to 2^V$ satisfying the following three axioms:

$$(t1) \quad u \in R(u, v) \text{ for all } u \text{ and } v \text{ in } V,$$

$$(t2) \quad R(u, v) = R(v, u) \text{ for all } u \text{ and } v \text{ in } V.$$

$$(t3) \quad R(u, u) = \{u\} \text{ for all } u \text{ in } V.$$

The term transit function is due to Mulder [22]. A transit function is also known in literature as interval function, for example, see Calder [2].

Let Φ be a property of paths in G, for instance the property of being a *geodesic*, i.e. a shortest path. A Φ-*path* is a path having property Φ. Formally, we take a *path property* Φ to be a subset of the set of all paths in G. Thus, if P is a Φ-path, then we may denote that also as $P \in \Phi$. Let u and v be vertices of G. Then $\Phi(u, v)$ denotes the subset of all u, v-paths in Φ. Here we consider *feasible* path properties, that is, path properties

Φ such that $\Phi(u, v) \neq \emptyset$, for all u, v in V. Any feasible path property Φ defines a transit function on G. Well-known instances are: the geodesics, the induced paths, and all paths. Transit function may be defined in terms of paths in the graph $G = (V, E)$. Such transit functions are called *path transit functions* on G in [22]. See also [7].

Given a transit function R on G, a subset W of V is *R-convex* if $R(u, v) \subseteq W$, for all u, v in W. A subgraph is convex if it is induced by a convex set. In general, a non-empty family $\mathcal{C}$ of subsets of a non-empty set V is called an abstract convexity on V if $\mathcal{C}$ is closed under intersections and nested unions, and both $\emptyset$ and V are elements of $\mathcal{C}$, and $(V, \mathcal{C})$ is known as a convexity space. It can be easily verified that the family $\mathcal{C}_R$ of all R-convex sets in V is an abstract convexity. In this paper, we consider mainly R-convexities defined by the induced and triangle-induced path transit functions.

Given a convexity space $(V, \mathcal{C})$, the smallest convex set containing a subset W of V is denoted by $\langle W\rangle_C$ and is called the *C-convex hull* of W. If $\mathcal{C}$ is an R-convexity, then the R-convex hull of W is denoted as $\langle W\rangle_R$.

If R is a transit function on $V(G)$, the iteration number of an S subset of V with respect to R is defined as follows:

$$R(S) = \cup\{R(u, v)|u, v \in S\}.$$

$$R^2(S) = \cup\{R(u, v)|u, v \in R(S)\}. \text{ etc.},$$

$$R^t(S) = \cup\{R(u, v)|u, v \in R^{(t-1)}(S)\}$$

Now the iteration number of S associated with the transit function R, denoted as *in(S)* is defined as the smallest number k such that $R^k(S) = R^{(k+1)}(S) = \langle S\rangle$. The iteration number of G associated with R is defined as the supremum of in(S) over all subsets of S of $(V(G)$.

A subset $S \subseteq V$ is called *convexly independent (independent)* if $x \notin \langle S - x\rangle_C$ for every $x \in S$.

The *rank* of $\mathcal{C}$ is the supremum of the cardinalities of the independent subsets of V. Clearly the Helly number is a lower bound for the rank.

The *hull number* hn_C of $\mathcal{C}$ is the minimum of the cardinalities of subsets S of V such that $\langle S\rangle_C = V$.

A clique in G is a set of pairwise adjacent vertices in G. For a vertex x in G, $N(x)$ denotes the neighbourhood of vertex x in G: $N(x) = \{y \in V(G) : y \neq x, xy \in E(G)\}$. For $W \subseteq V$, $N(W) = (\cup_{w \in W} N(w))\backslash W$.

The *clique number* ω is the cardinality of the largest clique in G.

For a subset A of V, the subgraph induced by A is denoted by $G(A)$. For $X \subseteq V$, $\mathcal{C}(X)$ denotes the set of connected components of $G(V\backslash X)$ (connected components as vertex sets). S is called a separator if $|\mathcal{C}(S)| \geq 2$, a clique separator if S is a separator and a clique, an *a-b* separator if a and b are in different components of $\mathcal{C}(S)$, a minimal *a-b* separator if S is an *a-b* separator and no proper subset of S is an *a-b* separator. S is called a minimal separator if there is some pair of vertices a, b such that S is a minimal *a-b* separator. This can be equivalently expressed as follows: S is a minimal separator in G if and only if there are two components $C_1, C_2 \in \mathcal{C}(S)$ such that $N(C_1) = N(C_2) = S$. C_1 and C_2 are then called the full components of $\mathcal{C}(S)$. Note that S is a minimal *a-b* separator if and only if a and b belong to two different full components of $\mathcal{C}(S)$. S is called a minimal clique separator if S is a clique and a minimal separator.

An *atom* of G is a maximal connected subgraph of G that has no clique separator and is maximal with respect to this property. We have the following remark.

Remark 1. Note that a subgraph $G(A)$ has no clique separator if and only if $G(A)$ has no minimal clique separator: as any separator contains some minimal separator. Thus if A has a clique separator, then it has a minimal clique separator.

Using this remark, we define the following intersection graph on G.

The *atom-clique separator tree* $T(G)$ of G is defined as follows. The vertices of $T(G)$ are the atoms and minimal clique separators of G. The edges of $T(G)$ are defined as follows: if A is an atom and S a minimal clique separator such that S separates vertices of A from vertices of $G(V\backslash(S \cup A))$ and $N(A) = S$. It can be verified that $T(G)$ is a tree. For if it has a cycle, say C, then there is a minimal clique separator in C the removal of which does not disconnect G, a contradiction.

As mentioned in the introduction, the algorithm for decomposition of a connected graph G by minimal clique separators is designed by Berry et.al. in [1]. Now we present the decomposition algorithm described in [1].

Algorithm 1 (Decomposition). Given a connected graph G, choose (if there is one), a minimal clique separator S in G, decompose G into the set of subgraphs $\{G(C_i \cup N(C_i))|C_i \in (S)\}$.

Remark 2. This decomposition can be represented by a decomposition tree, and each edge is labelled by a full component C of some clique separator $N(C)$, which is also a minimal clique separator of the initial graph G. At each step of the decomposition, the minimal clique separator S being processed disappears from the set of minimal separators of the components, as well as any minimal separator included in it. All other minimal separators are partitioned into the subgraphs obtained.

We illustrate the algorithm with an example.

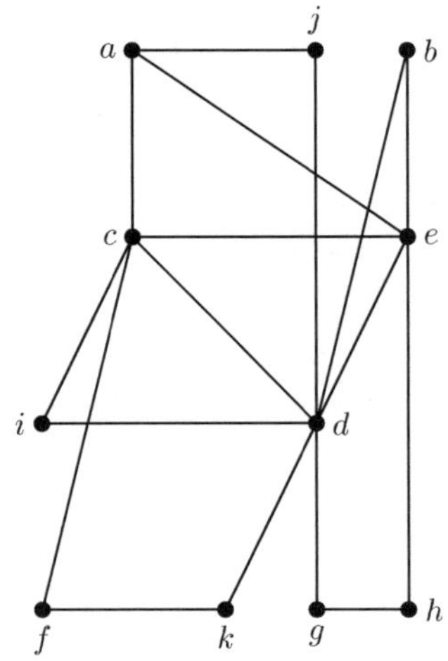

Graph G

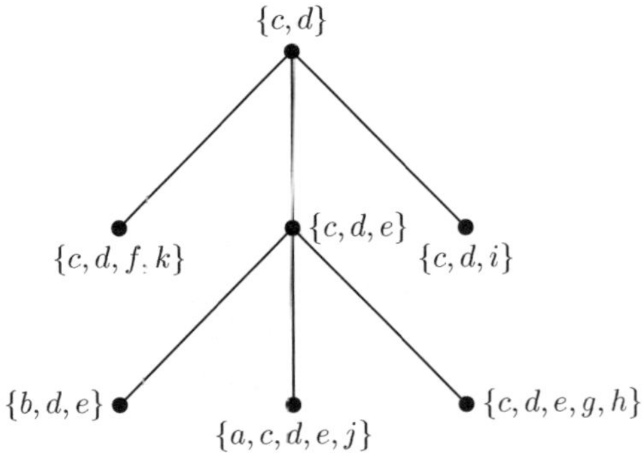

Decomposition Tree of G

Example 1. The Graph G has three minimal clique separators: $\{c, d\}$, $\{d, e\}$ and $\{c, d, e\}$. Step 1: minimal clique separator $\{c, d\}$ of G defines three connected components: $\{i\}$, $\{f, k\}$ and $\{a, b, e, g, h, j\}$. Atoms $\{c, d, i\}$ and $\{c, d, f, k\}$ are generated. Step 2: $\{c, d, e\}$ is a minimal clique separator of $G(\{a, b, c, d, e, g, h, j\})$, and also of G, and defines three connected components: $\{a, j\}$ and $\{g, h\}$, which are full, and $\{b\}$, which is non-full. Atoms $\{a, c, d, e, j\}$ and $\{c, d, e, g, h\}$ are generated by minimal separator $\{c, d, e\}$; b being non-full, we restrict the minimal separator used to $N(b) = \{d, e\}$: atom $\{b, d, e\}$ is generated. Note that minimal clique separator $\{d, e\}$ does not appear explicitly, as it is included in $\{c, d, e\}$ and processed at the time, by way of the non-full component $\{b\}$ of $\mathcal{C}(\{c, d, e\})$. A simple observation reveals that the *decomposition* tree of G obtained using the *Algorithm* 1 is the same as the *Atom-Clique separator* tree $T(G)$, which we state as the next theorem.

Theorem 1. *For a connected graph G, the atom-clique separator tree $T(G)$ of G is isomorphic to the decomposition tree obtained using Algorithm 1.*

Proof. If S is a minimal clique separator of G and C_i's are full components of S, then in the decomposition tree, we have the edges between S and C_i's. By definition of $T(G)$, we have edges between the vertices corresponding to S and C_i's. Hence the Theorem follows. □

3. The induced path convexity

The *induced path transit function* J_G of G is defined as follows:

$$J_G(u, v) = \{z \in V \mid z \text{ lies on some induced } u, v - \text{path in } G\}$$

for each $u, v \in V$.

The induced path convexity is the associated convexity of the J-transit function. The induced path convexity is also known as monophonic convexity and minimal path convexity, see [10], [24] and [17] respectively. Duchet has determined in [10] the relationships between the Helly and Radon numbers and the Clique number for this convexity. It is also shown there that the Carathéodory number c satisfies $c \leq 2$.

We state the characterization of the J-convex hull by Duchet [10], which is the basic fact used in developing an algorithm for finding the J-convex hull of a subset of $V(G)$.

Lemma 2 (Duchet). *In a connected graph G, a vertex 'w' belongs to the J-convex hull of A, ($\langle J(A)\rangle$) if and only if no clique of $G \backslash w$ separates 'w' and A.*

Also, note that in [15], a related notion to that of the hull number is the monophonic number in the case of the induced path convexity and geodetic number in the case of the geodesic convexity. The relation between these parameters has been considered by many authors, for example, [8], [13] and [14] for the geodesic convexity and [15] and [24] for the induced path convexity. We recall the definition of the monophonic set and monophonic number. Through out this section we denote the hull number of a graph G corresponding to the J-convexity by hn.

A subset W of $V(G)$ is called a monophonic set if $J(S) = V$. The monophonic number of G is the minimum cardinality of a monophonic set.

A graph G is said to be J-monotone if $J(u, v)$ is J-convex for every $u, v \in V$. J-monotone graphs have been studied in [3–5].

The iteration number of the J-transit function is called the minimal path iteration number in [3]. The following result is proved in [3].

> The induced path transit function J on a connected graph G is *monotone* if and only if the minimal path iteration number of G is one.

From this result we have that, if G is J-monotone, then for any subset W of $V(G)$, $\langle J(W) \rangle = J(W)$. Therefore, the following corollary holds.

Corollary 3. *If a connected graph G is J-monotone, then the hull number of G is the same as the monophonic number of G.*

The converse of this corollary doesn't hold, for example, the complete bipartite graph $K_{2,3}$ is not J-monotone and it can be easily verified that $K_{2,3}$ has both the hull number and monophonic number equal to 2.

If G is K_n, then the hull number hn is n since the vertex set of K_n is convexly independent and if G is itself an atom which is not a clique, then $hn = 2$, since any two non-adjacent vertices of G have the whole vertex set as their convex hull. Thus, we have the following observation.

Observation 4. *If G is the complete graph K_n, the hull number hn for the induced path convexity of G is n and if G is an atom, which is not a clique, then hn is 2.*

Let G be a connected graph which is neither complete nor an atom. Let $T(G)$ denote the atom-clique separator tree of G such that every leaf of $T(G)$ is not a clique. Let $L(T(G))$ denote the set of leaves of $T(G)$. We have the following proposition.

Proposition 5. *Let G be a connected graph which is not an atom or a complete graph K_n such that every leaf of $T(G)$ is not a clique. Then the Hull number hn for the induced path convexity of G is given by $hn = |L(T(G))|$.*

Proof. By the definition of $T(G)$, every leaf of $T(G)$ represents an atom of G and the atom corresponding to a leaf of $T(G)$ is not a clique. Corresponding to every element a' in $L(T(G))$, there exists an atom A of G and take a representative vertex, say a in A. Let $S \subseteq V$ consists of representative vertices from the atoms corresponding to every element in $L(T(G))$. Clearly $|S| = |L(T(G)|$. We have to show that any vertex x of V is in the convex hull of two vertices, say u and v of S which corresponds to two distinct atoms, say A and B respectively in $L(T(G)$. Also, we have to show that S is a set with minimum cardinality satisfying the above condition.

We first show the minimality of S. For this, we prove that every element a of S is in the hull set of G. Since a belongs to an atom of G, which corresponds to a leaf of $T(G)$, there exists a minimal clique separator separating a and $G \backslash a$. Therefore a does not belong to an induced path between any pair of vertices in $G \backslash a$ by using Lemma 2, which shows that $\langle J(V(G \backslash a)) \rangle$ does not contain vertices in A. Therefore a

must belong to the hull set of G. This shows that the set S is convexly independent and must belong to the hull set of G. Now, we prove that S is a hull set. Suppose not. That is, $\langle S \rangle \neq V$. Then there exists a vertex x of G such that $x \notin \langle S \rangle$. Then by Lemma 2, there exists a clique separator K separating x and $\langle S \rangle$. Using Remark 1, without loss of generality, K can be taken as a minimal clique separator. If $x \notin \langle S \cup K \rangle$, then again by Lemma 2 and Remark 1, we get another minimal clique separator K_1 separating x and $\langle S \cup K \rangle$. Continuing like this, we can get a sequence of minimal clique separators $K_1, K_2, \ldots, K_m$ such that $x \in \langle S \cup K \cup K_1 \cup \cdots \cup K_m \rangle$. If $\langle S \cup K \cup K_1 \cup \cdots \cup K_m \rangle \neq V$, then we can replace $S \cup K \cup K_1 \cup \cdots \cup K_m$ by S and there exists an x_1 such that $x_1 \notin \langle S \rangle$. Arguing as before, since G is finite, we obtain an x_n and a sequence of minimal clique separators $L_1, L_2, \ldots, L_r$ such that $x_n \notin \langle S \cup L_1 \cup L_2 \cup \cdots \cup L_{r-1} \rangle$ and $x_n \in \langle S \cup L_1 \cup L_2 \cup \cdots \cup L_r \rangle = V(G)$, which shows that x_n belongs to an atom that corresponds to a leaf of $T(G)$, which is a contradiction. Therefore the set S corresponding to $L(T(G)$ is a hull set. Thus S is a convex independent as well as a hull set and so is a minimal hull set and hence the hull number is exactly $|S| = |L(T(G)|$. □

Suppose one of the atom A corresponding to a leaf of $T(G)$ is itself a clique. Let S be the minimal clique separator separating A from the rest of G. If the clique $A \backslash S$ is of size $m \geq 2$, then add m new vertices to the graph $T(G)$ and join all of them by edges from the vertex corresponding to A in $T(G)$. Repeat this procedure for every clique atom corresponding to the leaf in $T(G)$ and the tree thus obtained is denoted by $T'(G)$. Denote the set of leaves of a tree $T'(G)$ by $L(T'(G))$. For each x in $L(T'(G))$ take a corresponding vertex in $V(G)$ as before and the set of vertices thus obtained forms a convex-independent set of $V(G)$ with respect to the induced path convexity. We have a straightforward proposition using Proposition 5.

Proposition 6. *Let G be a connected graph which is not an atom and different from K_n. For the induced path convexity of G, the Hull number hn satisfies $hn(G) = |L(T'(G))|$.*

Based on these propositions we give the following algorithm for computing the hull number and hull set of a graph. For a node X in the decomposition tree, let $P(X)$ denote the parent node of X.

Algorithm 2 (Algorithm Hull Set and Hull number).

Input: A Graph G

Output : The hull number of G and a hull set of G

step 0. Initialize hull number to 0 and hull set to $\emptyset$.

Step 1. Find a decomposition T by minimal clique separators of G

Step 2. For each leaf L of T do If L is a clique add $|L| - |P(L)|$ to the hull number and all vertices of $L \backslash P(L)$ to hull set
else
add 1 to the hull number and add any one vertex of L to the hull set

step3. Return hull number and hull set.

The time complexity of the above algorithm is also $O(mn)$ since the time complexity of decomposition is $O(mn)$ for an (m, n)-graph.

Theorem 7. *Algorithm 2 is correct.*

Proof. The correctness of the hull number follows from Proposition 6 and Theorem 1. The correctness of the hull set follows from the proof of Proposition 5 with the following observation. Since we are taking one representative vertex from each atom which is not a clique and the vertices of atoms which are cliques do not belong to the clique minimal separator, this is equivalent to taking a representative vertex from each of the leaves of $T'(G)$. Hence the Theorem. □

Clearly, every clique of G forms a convex-independent set in the induced path convexity of G. The cardinality of $L(T'(G))$ or $L(T(G))$ is dependent on the number of external atoms and not dependent on ω: there can be graphs with $\omega \geq hn$ and $hn \geq \omega$. Therefore we have the following corollary.

Corollary 8. *The rank of the J-convexity is given by max(ω, hn).*

Remark 3. Since finding the maximum clique problem in a graph is NP-complete, by Corollary 8, finding the rank of the induced path convexity is also NP-complete.

3.1 *The induced path convex hull algorithm*

A clique separator is *maximal* if it is not properly contained in another clique separator.

Components of a *clique separator* C: The components of a clique separator C are the components of $G \backslash C$. The induced path convex hull of subset A of $V(G)$ is denoted by $\langle J(A) \rangle$.

In this section, we give an algorithm for computing the induced path convex hull of a subset of vertices of a connected graph $G = (V, E)$. The algorithm is developed using Tarjan's decomposition algorithm of connected graphs by clique separators [30]. Tarjan's algorithm consists in repeatedly finding a clique separator, and copying this clique into each of the connected components thus defined. This implementation is an important tool for a divide and conquer approach to many hard graph problems.

The complexity of our algorithm is the same as that of the decomposition algorithm for connected graphs. The basic idea of the first algorithm is to apply the decomposition algorithm using the minimal and maximal clique separators.

Our algorithm is based on the characterization of the induced path convex hull of a subset of $V(G)$. Based on this characterization, our method is to find a clique separator and check whether there is a component of $G \backslash C$ not containing a vertex of A and eliminate these components. We continue the process until getting $\langle J(A) \rangle$.

An *elimination ordering* π is a numbering of the vertices of G from 1 to n. The fill-in F_π caused by the ordering π is the set of edges defined as follows: $F_\pi = \{\{v, w\} | v \neq w$, v not adjacent to w and there is a path $v = v_1, v_2, \ldots, v_k = w$ in G such that $\pi(v_i) < min\{\pi(v), \pi(w)\}$, for $i = 2, \ldots, k-1\}$.

An *elimination ordering* π is *minimum* if $|F_\pi|$ is minimum over all possible orderings, and *minimal* if there is no ordering σ such that $F_\sigma \subsetneq F_\pi$. The graph $G_\pi = (V, E \cup F_\pi)$ is the *fill-in* graph for π.

The decomposition *algorithm* due to Tarjan is the following [30].

> The algorithm consists of two steps. First, find a minimal ordering π and compute $C(v) = \{w|\pi(w) > \pi(v) \text{ and } \{v, w\} \in E \cup F_\pi\}$ for each vertex v. Repeat the following step for each vertex v in increasing order with respect to π.
>
> Decomposition step. Let A be the vertex set of the connected component of $G(V\backslash C(v))$ containing v, and let $B = V\backslash(C(v) \cup A)$. If $C(v)$ is a clique in G and $B \neq \emptyset$, decompose G into $G' = G(A \cup C(v))$ and $G'' = G(B \cup C(v))$, separated by $C(v)$. Replace G by G''.

A decomposition step is successful if it finds a clique separator. The running time of the algorithm is $O(mn)$. The time for computing $C(v)$ is also $O(mn)$. Tarjan, for proving the correctness of this algorithm, has established that every minimal clique separator is of the form $C(v)$ for some vertex v. But it may be noted that every $C(v)$ need not be a clique separator. Tarjan has also noted that the algorithm can be modified using decomposition by maximal clique with the same complexity $O(mn)$. Also the maximal cliques are either of the form $C(v) \cup \{v\}$ or $B' \cup C(v)$, where $C(v)$ is a minimal clique separator and $B' \subseteq B$, of the decomposition step. In our algorithm, we write $C'(v)$ for the maximal clique separator containing the minimal clique separator $C(v)$.

Now we state the algorithm for computing the induced path convex hull $\langle J(A)\rangle$ of a V-subset A. Reducing the graph G by G_1, we mean the graph obtained by removing the components of $G\backslash C$ not containing vertices of A, where C is a clique separator. We denote the induced subgraph of a subset A of vertices by $G(A)$ and by $\langle J(A)\rangle$, we mean the subgraph induced by $\langle J(A)\rangle$.

Algorithm 3.

Step 1: If A is a clique, then $\langle J(A)\rangle = G(A)$. Stop. Else,

Step 2: Set $X = \emptyset$. Initialization

Step 3: For each vertex $v \in G$, find $C(v)$, $C'(v)$

Step 4: Try to reduce G to G_1 by $C(v)$. If $G_1 \neq G$, go to Step 1 with $G = G_1$, else

Step 5: Try to reduce G to G_1 by $C'(v)$. If $G_1 \neq G$ go to Step 1 with $G = G_1$, else set $X = X \cup C(v)$ and go to Step 6.

Step 6: For each vertex $v \in G$, set $G_1 = G\backslash[C'(v)\backslash(X \cup A)]$, if $G \neq G_1$ go to Step 1 with $G = G_1$

Step 7: $\langle J(A)\rangle = G_1$

Theorem 9. *Algorithm 3 is correct.*

Proof. For a subset A of $V(G)$, let G_1 be the subgraph obtained by the algorithm. We prove that $\langle J(A)\rangle = G_1$. It is trivial to note that $\langle J(A)\rangle = G(A)$, if A is a clique and in that case we have $\langle J(A)\rangle = G_1$. Suppose $\langle J(A)\rangle = V(G)$. Then by Lemma 2, for any vertex $v \in G$, no clique separator of G will separate v and A. Therefore, every component of either a minimal clique separator or a maximal clique separator contains vertices of A. Therefore Step 4 and Step 5 of the algorithm will result in $G_1 = G$. Also after Step 7 of the algorithm, we get $G = G_1$ for any $v \in V(G)$, otherwise the clique defined in Step 6, namely $C'(v)\backslash(X \cup A)$ will separate at least one vertex from A. Therefore, we get that $G_1 = G$ and hence $G_1 = \langle J(A)\rangle$. Now suppose that $\langle J(A)\rangle \neq V(G)$. Therefore there exists a vertex $x \notin \langle J(A)\rangle$. Again by Lemma 2, there exists a clique separator C separating x and A. If C is a minimal clique separator of the form $C(v)$, for some v, then there exists a component of $G\backslash C(v)$ not containing vertices of A. Therefore by Step 4, we get that $x \notin G_1$. Similarly, if C is a maximal clique separator, by Step 5, we get that $x \notin G_1$. If C is neither a maximal nor a minimal clique separator, then there exists a minimal clique separator $C(v)$ and a maximal clique separator $C'(v)$ such that $C(v) \subsetneq C \subsetneq C'(v)$, for some v. Therefore, if neither $C(v)$ nor $C'(v)$ separates x and A, then $x \in [C'(v)\backslash(X \cup A)]$ by Step 6 and hence $x \notin G_1$. Therefore, we get $G_1 \subseteq \langle J(A)\rangle$.

Conversely, to prove that $\langle J(A)\rangle \subseteq G_1$. If $G_1 = G$, then all components of all minimal and maximal clique separators contain vertices of A. Also, since $G_1 = G$ after Steps 4 and 5, every clique C is such that $C(v) \subsetneq C \subsetneq C'(v)$, for some v, every component of C also contains vertices of A and hence no clique separator can separate A and any vertex of G. Therefore $\langle J(A)\rangle = G$.

If $G_1 \neq G$, then there exists vertex $x \notin G_1$. This can happen in either Step 4, 5 or 7. If $x \notin G_1$ after Step 4 or Step 5 or Step 7, then there exists a minimal clique separator $C(v)$, a maximal clique separator $C'(v)$, the clique separator $[C'(v)\backslash(X \cup A)]$ respectively, separating x and A, which by Lemma 2 shows that $x \notin \langle J(A)\rangle$, which completes the proof. □

Since we are applying the decomposition algorithm of Tarjan, the complexity of the algorithm is the same as that of finding $C(v)$ and $C'(v)$ which is bounded by $O(mn)$.

4. Triangle-path convexities

In this section, we analyse the Hull number for the triangle path convexity. We formally define a 'triangle path' in a graph. Let $P = u_1 \to u_2 \to \cdots \to u_k$ be a path in G. Let z_i be a vertex not on P but adjacent to two consecutive vertices u_i, u_{i+1} of P. Then we say that the path $Q = u_1 \to u_2 \to \cdots \to u_i \to z_i \to u_{i+1} \to \cdots \to u_k$ is obtained from P by replacing the edge $u_i \to u_{i+1}$ by a triangle. A *triangular extension* of a path P is a path Q obtained from P by replacing some of the edges of P by triangles. We will call P a triangular extension of itself as well. Let Φ be a (feasible) path property on G. Then $\Phi^{\triangle}$ is the path property defined by

$$\Phi^{\triangle} = \{Q \mid Q \text{ is a triangular extension of some path in } \Phi\}.$$

Note that we have $\Phi \subseteq \Phi^{\triangle}$, with equality if and only if no path in Φ is involved in a triangle. In particular, we have equality in the case of a triangle-free graph.

Let R be a Φ-path transit function on G. Then $R^{\triangle}$ is the path transit function on G defined by

$$R^{\triangle} = R_{\Phi^{\triangle}}.$$

The transit function $R^{\triangle}$ is a *triangle path transit function*. Note that $R^{\triangle} = R$ if no path in Φ is involved in a triangle. If we take Φ as 'geodesics' or 'induced', we get the 'triangle geodesic transit function' or 'triangle induced path transit function'. The well studied triangle path transit function is the 'triangle induced path transit function' denoted as '$J^{\triangle}$', [3,7].

A characterization of the $J^{\triangle}$-convex hull is obtained in [3] as follows:

> Let $G = (V, E)$ be a connected graph, and let $W \subseteq V$. Then a vertex v does not belong to the $J^{\triangle}$-convex hull of W if and only if there exists a clique M separating v and W in such a way that any two paths connecting v to two distinct vertices of M contain a chordless cycle of length at least 4.

In order to discuss the hull number and rank, we need a small modification for the atom-minimal clique separator $T(G)$ of G in case of the $J^{\triangle}$-convexity, denoted by $T^{\triangle}(G)$. The vertex set of $T^{\triangle}(G)$ is the family of all atoms and minimal clique separators not separating a $J^{\triangle}$-convex subset, and an atom A is adjacent to a minimal clique separator M in $T^{\triangle}(G)$ if M separates A such that any two paths connecting a vertex in A to two distinct vertices of M contain a chordless cycle of length at least equal to 4. Since we are not considering all minimal clique separators, clearly $T^{\triangle}(G)$ will be a subtree of $T(G)$ and as in the case of $L(T(G))$, the subset of V corresponding to the set of elements of $L(T^{\triangle}(G))$, forms a convex-independent set of $V(G)$. Again we have a straightforward result similar to that of the induced path convexity.

Proposition 10. *The hull number of $J^{\triangle}$-convexity is equal to the cardinality of $L(T^{\triangle}(G))$.*

Unlike in the J-convexity, cliques of G are not $J^{\triangle}$-convex-independent. Hence we have

Proposition 11. *The rank of the $J^{\triangle}$-convexity is the hull number.*

4.1 *The all-paths convexity*

The *all-paths transit function* A_G of G is defined as:

$$A_G(u, v) = \{x \in V \mid x \text{ lies on some } u, v\text{-path in } G\},$$

for u, v in V.

The convexity associated with A is the 'all-paths convexity' which is the coarsest path convexity. The all-paths transit function is characterized using axioms in [6].

By considering the block-cut-vertex tree $B(G)$ instead of the $T(G)$, and following similar arguments to that in the case of the $J^{\triangle}$-convexity, we can obtain the hull number hn and the rank of the A-convexity. We summarize these results in the following remark.

Remark 4. The following statement holds for the all-paths convexity for a connected graph G. The hull number and the rank are both equal to the number of leaves of the block-cut-vertex tree $(B(G))$ of G.

References

[1] A. Bery and J. P. Bordat, Decomposition by clique minimal separators. Research report, LIM, Marseiller (1997)

[2] J. Calder, Some elementary properties of interval convexities, *J. London Math. Soc.*, **3** (1971) 422–428.

[3] M. Changat and J. Mathew, On triangle path convexity in graphs, *Discrete Math.*, **206** (1999) 91–95.

[4] M. Changat and J. Mathew, Induced path transit function, monotone and Peano axioms. Interval monotone graphs, *Discrete Math.* **286** (2004) 185–194.

[5] M. Changat and J. Mathew, Characterizations of J-monotone graphs (in this volume) 47–55.

[6] M. Changat, S. Klavžar and H. M. Mulder, The all-paths transit function of a graph, *Czech. Math. J.*, **51**, (2001) no. 126, 439–448.

[7] M. Changat, H. M. Mulder and G. Sierksma, Convexities related to path properties on graphs, *Discrete Math.*, **290** (2005) 117–131.

[8] G. Chartrand, F. Harary and P. Zhang, On the hull number of a graph, *Ars Combinatoria.*, **57** (2000) 129–138.

[9] P. Duchet, Convexity in combinatorial structures, *Rend. Circ. Mat. Palermo (2) Suppl.*, **14** (1987) 261–293.

[10] P. Duchet, Convex sets in graphs II. Minimal path convexity, *J. Combin. Theory Ser. B.*, **44** (1988) 307–316.

[11] P. Duchet, Discrete convexity: retractions, morphisms and partition problem. Proc. of the conf. on graph connections, (eds.) R. Balakrishnan, H. M. Mulder and A. Vijayakumar, New Delhi (1999) 10–18.

[12] J. Eckhoff, Helly, Radon and Carathéodory type theorems, in: Handbook of convex geometry. (eds.) P. H. Gruber, J. W. Wills, North-Holland (1993) 389–448.

[13] M. G. Everett and S. B. Siedman, The hull number of a graph, *Disc. Math.*, **57** (1985) 217–223.

[14] F. Harary, E. Loukakis and C. Tsousros, The geodetic number of a graph, *Math. Computer modelling*, **17** (1993) no. 11, 89–95.

[15] C. Hernando, M. Mora, I. M. Pelayo and C. Seara, On monophonic sets in graphs (preprint).

[16] R. E. Jamison, Partition numbers for trees and ordered sets, *Pacific J. Math.*, **96** (1981) 115–140.

[17] R. E. Jamison-Waldner, A perspective on abstract convexity: classifying alignments by varieties, in convexity and related combinatorial geometry. Proceedings of 2nd Oklahoma conf., (eds.) D. C. Kay and M. Breen, New York (1982) 113–150.

[18] S. Klavžar and H. M. Mulder, Median graphs; characterizations, location theory and related structures, *J. Combin. Math. Combin. Comp.*, **30** (1999) 103–127.

[19] K. Kolodziejczyk and G. Sierksma, The semirank and Eckhoff's conjecture for Radon numbers, *Bull. Soc. Math. Belg.* **42** (1990) 383–388.

[20] M. A. Morgana and H. M. Mulder, The induced path convexity, betweenness and svelte graphs, *Disc. Math.*, **254** (2002) 349–370.

[21] H. M. Mulder, The interval function of a graph, Mathematical Centre Tracts 132, Mathematisch Centrum, Amsterdam (1980).

[22] H. M. Mulder, Transit functions on graphs (and posets), (in this volume) 117–130.

[23] L. Nebeský, A characterization of the interval function of a connected graph, *Czech. Math. J.*, **44** (1994) no. 119, 173–178.

[24] I. M. Pelayo, Generalizing the Krein-Milman property in graph convexity spaces: a short survey, (in this volume) 131–142.

[25] E. Sampathkumar, Convex sets in graphs, *Indian J. Pure Appl. Math.*, **15** (1984) 1065–1071.

[26] G. Sierksma, Carathéodory and Helly numbers of convex product structures, *Pacific. J. Math.*, **61** (1975) 275–282.
[27] G. Sierksma, Axiomatic convexity theory and convex product space, Ph.D. Thesis, Univ. of Groningen (1976).
[28] G. Sierksma, Relationships between Carathéodory, Helly, Radon and Exchange numbers of convexity spaces, *Nieuw Archief voor Wiskunde*, **3** (1977) no. 25, 115–132.
[29] M. L. J. van de Vel, Theory of convex structures, North Holland, Amsterdam (1993).
[30] R. E. Tarjan, Decomposition by clique separators, *Disc. Math.*, **55** (1985) 221–232.

Joint Proc. Int. Wrk.shps. Conv. in Discr. Struct. and Metric & Conv. Graph Th. No. 5, 2008, pp. 25–46.

Discrete Metric Spaces and Related Graph Classes

Manoj Changat[1] and Victor Chepoi[2]

[1]Department of Futures Studies, University of Kerala, Thiruvananthapuram 695 034, India
[2]Laboratoire d'Informatique Fondamentale de Marseille, Université de la Méditerranée, Faculté des Sciences de Luminy, F-13288 Marseille Cedex 9, France
e-mail: mchangat@gmail.com; chepoi@lif.univ-mrs.fr

Abstract. The purpose of these lecture notes is to survey the most important classical and recent results about finite metric spaces, their embedding and related graph classes.

1. Introduction

1.1 *Avant-propos*

The study of finite metric spaces and metric embeddings has its origins in Banach space theory and functional analysis, having substantial connections with finite dimensional normed spaces. In recent years, this area has emerged as a new and influential branch of discrete mathematics, with deep and surprising applications in computer science. New insights into metric embeddings have been developed as well as new applications and connections discovered to approximation algorithms, algorithms for large data sets, and algorithms for high dimensional computational geometry. These notes are based on lectures given by the second author at the International Workshop "Convexity in Discrete Structures" at Kovalam, Thiruvananthapuram in 2006. They should be considered only as lecture notes and not as a survey of all results in this area.

1.2 *Related references*

There exists a vast recent literature on finite metric spaces, isometric embeddings and distances in graphs. The book by Deza and Laurent [38] is our basic reference on distance geometry and exact isometric embeddings. The survey on metric graph theory given by Bandelt and Chepoi [12] is related with the second part of our lecture notes (Section 4). The study of low-distortion embeddings of graphs and finite metric spaces into l_2 or l_1-spaces, with numerous applications in the design of approximation algorithms, was initiated by Linial, London, and Rabinovich [61] and is the subject of several surveys or book chapters. Linial gave a survey on finite metric spaces and their connection to combinatorics, geometry, and algorithms at the International Congress of Mathematicians in 2002. Matousek, in his book "Lectures on Geometry", discusses the mathematical aspects of low-distortion embeddings [64]; see also [50] for a more complete treatment of the subject and the survey by Indyk [49] for algorithmic applications of low distortion embeddings.

1.3 *Basic definitions*

A *metric space* is a pair (X, d), where X is a non-empty set and a map $d : X \times X \to \mathbb{R}^+$, called the distance function, such that

(i) $d(x, y) = d(y, x)$,
(ii) $d(x, y) = 0$ iff $x = y$,
(iii) $d(x, y) \leq d(x, z) + d(z, y)$.

Basic examples of metrics are normed (alias Minkowski) metric spaces $(\mathbb{R}^m, d_{\|\cdot\|})$ in which the metric is derived from a norm $d_{\|\cdot\|}(x, y) = \|x - y\|$ for all $x, y \in \mathbb{R}^m$. In particular, the l_p-metrics for $p \geq 1$ are defined as $d_{l_p}(x, y) = \left[\sum_{k=1}^m |x_k - y_k|^p\right]^{\frac{1}{p}}$, where $x = (x_1, x_2, \ldots, x_m)$ and $y = (y_1, y_2, \ldots, y_m)$. Denote by l_p^m the metric space $(\mathbb{R}^m, d_{l_p})$. We have three basic host l_p-metric spaces: the Euclidean space l_2^m with $d_{l_2}(x, y) = \left[\sum_{k=1}^m |x_k - y_k|^2\right]^{\frac{1}{2}}$, the rectilinear (or l_1-)space l_1^m with $d_{l_1}(x, y) = \sum_{k=1}^m |x_k - y_k|$ and the space l_∞^m with $d_{l_\infty}(x, y) = \max\{|x_k - y_k| : 1 \leq k \leq m\}$.

Another example of a metric is the so-called Hamming distance d_H defined by setting $d_H(x, y) = |\{i \in \{1, \ldots, m\} : x_i \neq y_i\}|$ for all $x, y \in \mathbb{R}^m$. For binary vectors $x, y \in \{0, 1\}^m$, the Hamming distance $d_H(x, y)$, the l_1-distance $d_{l_1}(x, y)$ and the graph distance $d(x, y)$ of the n-cube coincide. Finally, the standard graph distance d_G also induces a metric space. Let $G = (V, E)$ be a connected, not necessarily finite, undirected and unweighted graph. Then the length of a path is the number of edges in this path. The distance $d_G(x, y)$ is the length of a shortest path connecting two vertices x, y of G.

If our metric space is finite, say it has n points, then we will denote it by (V_n, d), where $V_n = \{1, \ldots, n\}$ for convenience, and d is a metric on V_n. The distance matrix of such a space is the $n \times n$ symmetric matrix D whose (i, j)-th entry is $d(i, j)$ for all $i, j \in V_n$. Set $E_n = \{ij : i, j \in V_n, i \neq j\}$; then any distance d on V_n can be viewed as a vector $(d_{ij})_{1 \leq i < j \leq n} \in \mathbb{R}^{E_n}$. The triangle inequalities $d(i, j) + d(i, k) \geq d(k, j)$ for all $i, j, k \in V_n$ define a *convex cone* in the space $\mathbb{R}^{E_n}$, called the semimetric cone and denoted by MET_n.

We conclude with an important class of finite metrics: the planar metrics and the tree metrics. These are the metric spaces (V_n, d) such that one can construct a planar graph $G = (V, E)$ or a tree $T = (V, E)$ such that $V_n \subseteq V$ and $d = d_G|_{V_n}$.

1.4 *The structure of the lecture notes*

In Section 2, we describe isometric embeddings of arbitrary metric spaces into l_2, l_∞, and l_1-spaces. In Section 3, approximate (low distortion) isometric embedding into l_1 and l_2-spaces are discussed. In Section 4, we survey graph classes defined by distance properties.

2. Isometric embeddings

2.1 *Basic notions*

A metric space (X, d) is *isometrically embeddable* into a metric space (X', d') if there exists a mapping φ (the isometric embedding) from X to X' such that $d(x, y) = d'(\varphi(x), \varphi(y))$ for all $x, y \in X$. Then (X, d) is said to be an *(isometric) subspace* of (X', d'). The space (X, d) is l_p-embeddable if (X, d) is isometrically embeddable into the space l_p^m for some $m \geq 1$. The host space (X', d') is said to have order of congruence p if every metric space (X, d) embeds into (X', d') whenever each subspace Y of X with at most p points embeds into (X', d') and p is the smallest such integer (possibly infinite).

A metric space (X, d) is *distortion* β *embeddable* into a metric space (X', d') if there exists a mapping φ (the β-embedding) from X to X' such that $d(x, y) \leq d'(\varphi(x), \varphi(y)) \leq \beta d(x, y)$, for all $x, y \in X$ (or, equivalently, for l_p-embeddings: if there exist a mapping $\varphi : X \to X'$ such that $\frac{1}{\beta} d(x, y) \leq d'(\varphi(x), \varphi(y)) \leq d(x, y)$ for all $x, y \in X$.)

Further, we will investigate the following basic questions:

Questions. For a given host space or class of host spaces:

(i) is the order of congruence finite?
(ii) is the order of congruence bounded by a constant?
(iii) can the isometric subspaces be effectively characterized?
(iv) is the decision question *Is an input finite metric space isometrically embeddable into the host space* polynomial or NP-complete?
(v) find small distortion embeddings of finite metric spaces into the host space(s).
(vi) find small distortion embeddings into host spaces of small dimension.

2.2 *Isometric embeddings into trees*

The next theorem characterizes when a finite metric space isometrically embeds into a weighted tree.

Theorem 2.1 (Zaretsky, 1965; Buneman, 1971). *For a finite metric space (V_n, d) the following conditions are equivalent:*

(i) *(V_n, d) is a tree metric;*
(ii) *every quadruplet of V_n isometrically embeds into a tree;*
(iii) *(V_n, d) satisfies the four-point condition*

$$d(i, j) + d(l, k) \leq \max\{d(i, l) + d(j, k), d(i, k) + d(j, l)\}.$$

2.3 *Isometric embeddings into l_p-spaces, $p \neq 1, \infty$*

We continue with isometric embeddings into l_p-spaces.

Theorem 2.2 (Bretognolle, Dacunha-Castelle, Krivine, 1966). *A metric space (X, d) is l_p-embeddable if and only if every finite subspace of (X, d) is l_p-embeddable.*

Theorem 2.3 (Malitz & Malitz, 1992). *Let $p, m \geq 1$ be integers. Then a metric space (X, d) is l_p^m-embeddable if and only if every finite subspace of (X, d) is l_p^m-embeddable.*

Theorem 2.4 (Frechet). *Any n point metric space (V_n, d) can be isometrically embedded into l_∞^n.*

The isometric embedding φ in this theorem is constructed in the following way. For each point $i \in V_n$ define a coordinate $\varphi_i : V_n \to R_+$ by setting $\varphi_i(j) = d(i, j)$ and for each point $j \in V_n$ set $\varphi(j) = (\varphi_1(j), \ldots, \varphi_n(j))$.

Theorem 2.5 (Menger, 1923). *Given $m \geq 1$, a metric space (X, d) can be isometrically embedded in l_2^m if and only if (Y, d) embeds for every $Y \subseteq X$, $|Y| \leq m + 3$.*

To prove this theorem, a stronger assertion is established: (X, d) can be embedded into l_2^m but not into l_2^{m-1} if and only if there exists a subset $Y = \{x_0, x_1, \ldots, x_m\}$ of X such that

(i) (Y, d) can be embedded into l_2^m but not in l_2^{m-1};
(ii) for every $x, y \in X$, the metric space $(Y \cup \{x, y\}, d)$ can be isometrically embedded in l_2^m. The isometric embedding of Y into l_2^m then can be extended in a unique way to an isometric embedding of the whole space X into l_2^m.

Given a finite metric space (V_n, d), one can ask if it is possible to test in polynomial time if (V_n, d) embeds in l_2 (notice that Menger's theorem provides the answer only for a fixed dimension m). The answer is positive. For this, let φ be a mapping from V_n to l_2. Let $\varphi(i) = v_i$ and assume that $\varphi(1) = v_1 = \vec{0}$. Then $\|v_i - v_j\|^2 = d_{ij}^2$ for all $i, j \in V_n$ if and only if $\|v_i\|^2 + \|v_j\|^2 - 2\langle v_i, v_j\rangle = d_{ij}^2$, yielding $\langle v_i, v_j\rangle = \frac{1}{2}(d_{1i}^2 + d_{1j}^2 - d_{ij}^2)$, because $\|v_i\|^2 = \|v_i - v_1\|^2 = d_{1i}^2$, $\|v_j\|^2 = \|v_j - v_1\|^2 = d_{1j}^2$. Denote $A_{ij} = \frac{1}{2}(d_{1i}^2 + d_{1j}^2 - d_{ij}^2)$ and consider the matrix A. Then φ is an isometric embedding in l_2 if and only if $A_{ij} = \langle v_i, v_j\rangle$. That is, A can be written as $B^T B$, or, in other words, A is a positive semi-definite matrix (Schoenberg, 1932). Recognition of positive semi-definiteness can be done in polynomial time using an algorithm based on Gaussian elimination.

2.4 *Isometric embeddings into l_1-spaces*

We start with the following complexity result:

Theorem 2.6 (Karzanov, 1986). *Deciding if a finite metric space (V, d) l_1-embeds is NP-complete.*

Theorem 2.7. *For a finite metric space (V_n, d) the following conditions are equivalent:*

(i) *(V_n, d) is l_1-embeddable, that is there exists n vectors $u_1, \ldots, u_n \in R^m$ for some m such that $\|u_i - u_j\|_1 = d_{ij}, \forall i, j \in V_n$;*
(ii) *$d = \sum_{S \subseteq V_n} \lambda_S \delta(S)$ for some non negative λ_S, that is, d belongs to the cut cone $CUT_n = \left\{ \sum_{S \subseteq V_n} \lambda_S \delta(S) : \lambda_S \geq 0 \text{ for all } S \subseteq V_n \right\}$*

(iii) *there exist a measure space* $(\Omega, \mathcal{A}, \mu)$ *and events* $A_1, \ldots, A_n \in \mathcal{A}$ *such that* $d_{ij} = \mu(A_i \Delta A_j), \forall i, j \in V_n$.

In case of embedding into hypercubes or Hamming metric spaces, all λ_S in Theorem 2.7(i) are non negative integers and $d_{ij} = |A_i \Delta A_j|$. Ball (1990) established that the dimension m of $\mathbb{R}^m$ in Theorem 2.7(i) can be as large as $\frac{(n-3)(n-2)}{2}$ for $n \geq 4$. Fichet (1982) showed that $m \leq \frac{n(n-1)}{2} - 1$. Notice also that the difficulty to find an l_1-embedding via Theorem 2.7(ii) consists in finding the cuts (splits) $(S, V_n - S)$ such that $\lambda_S > 0$ (by Caratheodory's theorem, their number is $\leq \frac{n(n-1)}{2}$).

Theorem 2.8 (Bandelt & Chepoi, 1995). *A metric space* (X, d) *isometrically embeds into* $(\mathbb{R}^2, d_{l_1})$ *if and only if* (Y, d) *embeds for every* $Y \subseteq X$, $|Y| \leq 6$.

From Theorem 2.8 follows that the congruence orders of l_1^2 and l_∞^2 are equal to 6. Bandelt, Chepoi, and Laurent (1997) established that the congruence order of l_1^m is at least m^2 for $m \geq 3$ odd, at least $m^2 - 1$ for $m \geq 4$ even, and at least 10 for l_1^3. On the other hand, recently Edmonds (2006) established that the congruence order of l_∞^3 is not bounded.

Open Question. Find the congruence order of l_1^3. Is it finite? (It is only known to be ≥ 10).

Open Question. What is the largest set of an equilateral set of l_1^m? It is conjectured to be $2m$ but it is known to be $\leq m \log_2 m$ (Alon & Pudlak, 2003).

2.5 *Examples of l_1-embeddable spaces*

We continue with several important classes of l_1-embeddable metric spaces.

Example 2.9 (Tree metrics). Any tree with m leaves isometrically embeds into $l_1^{\frac{m}{2}}$.

Example 2.10 (Spherical metrics). Let $\mathbb{S}_m = \{x \in \mathbb{R}^{m+1} : \sum_{i=1}^{m+1} x_i^2 = 1\}$ be the m dimensional unit sphere and $d_{\mathbb{S}_m}(x, y) = arrcos(x^T y)$, for all $x, y \in \mathbb{S}_m$, denote the spherical distance, where $d_{\mathbb{S}_m}(x, y)$ is the geodesic distance on the sphere $\mathbb{S}_m$ between the points x and y (great circle metric). For $x \in \mathbb{S}_m$, let $H(x) = \{y \in \mathbb{S}_m : d_{\mathbb{S}_m}(x, y) \leq \frac{\pi}{2}\}$ be the hemisphere containing x. Consider the measure $\mu(A) = \frac{vol(A)}{vol(\mathbb{S}_m)}$ for $A \subseteq \mathbb{S}_m$.

Theorem 2.11 (Kelly, 1970). $\mu(H(x)\Delta H(y)) = \frac{1}{\pi} arrcos(x^T y) = \frac{1}{\pi} d_{\mathbb{S}_m}(x, y)$ *for all* $x, y \in \mathbb{S}_m$.

Example 2.12. l_2-Distances.

Theorem 2.13 (Schoenberg, 1935; Kelly, 1975). *For a finite metric space* (V_n, d), *if* d *is* l_2*-embeddable, then* d *is* l_1*-embeddable.*

To establish this result, first an l_2-embeddable metric space V_n is embedded into a sphere and then Theorem 2.11 is applied. The embedding into a sphere is obtained in the following way. Let $u_1, \ldots, u_n \in \mathbb{R}^m$, $d_{ij} = \|u_i - u_j\|_2$ for all i, j. Let $\mathbb{S}_{m,r}$ be the sphere of $\mathbb{R}^{m+1}$ of radius r and center $c = (0, \ldots, 0, r)$. Lift every point $u \in \mathbb{R}^m$ with $\|u\|_2 \leq r$

to a point $u^{(r)} \in \mathbb{S}_{m,r}$ by setting $u^{(r)} = (u, r - \sqrt{r^2 - (\|u\|_2)^2})$. Let $r \geq \max_{i=1}^{n} \|u_i\|_2$. Set $d^{(r)}(i, j) = r \cdot arccos(\frac{(u_i-c)^T(u_j-c)}{r^2})$, the spherical distance in $\mathbb{S}_{m,r}$ between the points $u_i^{(r)}$ and $u_j^{(r)}$. Then it can be shown that $\lim_{r\to\infty} d^{(r)}(i, j) = \|u_i - u_j\|_2$.

Example 2.14. Projective metrics in the plane.

A continuous metric d on $\mathbb{R}^2$ is called a *projective metric* if it satisfies $d(x, z) = d(x, y) + d(y, z)$ for any collinear points x, y, z lying in that order on a common line. The following theorem proved independently by Alexander in 1978 and Ambartzumian in 1977 gives a simple solution to the Hilbert's fourth problem in the plane.

Theorem 2.15 (Alexander, 1978; Ambartzumian, 1977). *Let d be a projective metric on $\mathbb{R}^2$. Then $(\mathbb{R}^2, d)$ is L_1-embeddable, namely there exists a positive Borel measure μ on the lines of $\mathbb{R}^2$ satisfying $2d(x, y) = \mu([[x, y]])$ for $x,y \in \mathbb{R}^2$, where $[[x, y]]$ denotes the set of lines crossing the segment $[x, y]$.*

The main step in the proof is to l_1-embed finite subspaces $P = \{p_1, \dots, p_n\}$ of the projective metric space $(\mathbb{R}^2, d)$. For this, Theorem 2.7 is used; namely, a measure on the lines crossing the segments $[p_i, p_j]$ with $p_i, p_j \in P$ is defined in the following way. For any line l, denote by K the (usual) convex hull of the points of P located left from l, and by K^* the convex hull of the points of P located right from l. Let $\sigma(K, K^*) = d(p, p^*) + d(q, q^*) - d(T) - d(T^*)$, where $d(T)$ and $d(T^*)$ are the perimeters (with respect to d) of the chains T and T^*. Then, using a nice double counting argument, it can be shown that for all $p_i, p_j \in P$, the following formula holds:

$$2d(p_i, p_j) = \sum\{d(e) : l \cap [p_i, p_j] \neq \phi\}.$$

Example 2.16. Kalmanson distances.

A distance d on V_n is called a Kalmanson distance if there exists a (circular) ordering $1, \dots, n$ of the points of V_n such that $\max\{d_{ij} + d_{rs}, d_{is} + d_{jr}\} \leq d_{ir} + d_{js}$ for all $i < j < r < s$ in the circular order.

Theorem 2.17 (Chepoi & Fichet, 1996). *A distance d is a Kalmanson distance if and only if it is circular decomposable. In particular, Kalmanson distances are l_1-embeddable.*

To establish this result, number the points of V_n in arbitrary way in circular order. For four points $i, i+1, j, j+1$ in this order, set $\delta_{ij} = d_{ij+1} + d_{i+1j} - d_{ij} - d_{i+1j+1}$. Notice that $\delta_{ij} \geq 0$ for any circular order compatible with a Kalmanson distance d, but, in general, we can have $\delta_{ij} < 0$. Now, the result follows from the following assertion: *For any finite metric space (V_n, d) and any circular ordering of V_n, we have $2d(u, v) = \sum\{\delta_{ij} : \{S_{ij}, \bar{S}_{ij}\}$ separates u and $v\}$ for all $u, v \in V_n$.*

Example 2.18. Totally decomposable metrics.

Recall that a metric d is l_1-embeddable if and only if $d = \sum_{S\subseteq V_n} \lambda_S \delta(S)$. In general, (i) it is difficult to find $\lambda_S \geq 0$ for all $S \subseteq V_n$, (ii) the l_1-decomposition is not unique, and (iii) there is an exponential number of cuts $(S, \bar{S})$ participating in the decomposition. To deal with this kind of problems, Bandelt & Dress (1992) introduced a *canonical decomposition* of every finite metric into a sum of $O(n^2)$ cut (split) metrics plus a residue. For a cut (split) $\{S, \bar{S}\}$ of V_n define its *isolation index* $\delta(S)$ by setting

$$\delta(S) = \frac{1}{2} \min_{a,a' \in S \ b,b' \in \bar{S}} \{\max\{d(a,b) + d(a',b'), d(a,b') + d(a',b), d(a,a') + d(b,b')\} - d(a,a') - d(b,b'\}.$$

Then $\{S, \bar{S}\}$ is called a *d-cut* if $\delta(S) > 0$ and denote by $\mathcal{S}$ the collection of all d-cuts.

Theorem 2.19 (Bandelt & Dress, 1992). *For any metric d on V_n, we have $d = \sum_{\{S,\bar{S}\} \in \mathcal{E}} \lambda_S \delta(S) + d'$, where d' is the prime residue.*

The metric d on V_n is called *totally decomposable* if $d' = 0$, that is, if $d = \sum_{\{S,\bar{S}\} \in \mathcal{S}} \lambda_S \delta(S)$. Bandelt & Dress (1992) established the following properties of totally decomposable metrics and d-cuts:

(i) there exists only $O(n^2)$ d-cuts (in fact, their (0,1)-vectors are linearly independent);
(ii) the d-cuts can be constructed in polynomial time;
(iii) the d-cuts are *weakly compatible*, that is, for any three d-cuts $\{S_1, \bar{S}_1\}$, $\{S_2, \bar{S}_2\}$, $\{S_3, \bar{S}_3\}$, $S_1 \cap S_2 \cap S_3 \neq \emptyset$ implies $\bar{S}_1 \cap \bar{S}_2 \cap \bar{S}_3 = \bar{S}_i \cap \bar{S}_j$ for some i, j.

Theorem 2.20 (Bandelt & Dress, 1992). *A finite metric space (V_n, d) is totally decomposable if and only if every 5-point subspace of V_n is totally decomposable.*

Examples of totally decomposable metrics are Kalmanson metrics, tree metrics, and Cartesian products of two trees, metric of some plane graphs, etc. Total decomposability of finite subspaces of (R^2, d_{l_1}) together with Theorem 2.19 play an important role in the proof of Theorem 2.8.

3. Low-distortion embedding of finite metric spaces into l_1, l_2 and l_∞

3.1 *Basic notions*

Recall that a mapping $\varphi : X \to X'$, where (X, d) and (X', d') are metric spaces, has distortion at most β, or is called a β-embedding, where $\beta \geq 1$, if there is a $c \in (0, \infty)$ such that for all $x, y \in X$, $c \cdot d(x, y) \leq d'(\varphi(x), \varphi(y)) \leq c \cdot \beta d(x, y)$. If X' is a normed space, we usually require $c = \frac{1}{\beta}$ or $c = 1$. If $c = 1$, we obtain $d(x, y) \leq d'(\varphi(x), \varphi(y)) \leq \beta d(x, y)$ (dilation). If $c = \frac{1}{\beta}$, we obtain $\frac{1}{\beta} d(x, y) \leq d'(\varphi(x), \varphi(y)) \leq d(x, y)$ (contraction). Mappings with bounded distortion are also called *bi-Lipschitz mappings*.

In this section, we will present the following results: (i) Bourgain's, Matousek's and Rao's low distortion embedding theorems; (ii) the Johnson-Lindenstrauss flattering lemma; (iii) probabilistic embeddings into tree metrics.

3.2 *Basic results*

We start with one of the principal results in this area proven by Bourgain in 1985.

Theorem 3.1 (Bourgain, 1985). *Every n-point metric space (V_n, d) can be embedded into l_1^m (in fact, into every l_p) with distortion $O(\log n)$ and dimension $m = O(\log^2 n)$.*

The embedding is constructed in the following way. Set $l = \log_2 n$ and $q = c \log_2 n$ (c is a suitable constant). Consider an embedding into l_1^{lq} with coordinates indexed by $i = 1, \ldots, l$ and $j = 1, \ldots, q$: set each $x \in V_n$ into A_{ij} with probability $\frac{1}{2j}$, all random choices being mutually independent. Set $\varphi(x)_{ij} = d(x, A_{ij})$, where $d(x, A_{ij}) = \min\{d(x, a) : a \in A_{ij}\}$ and $\varphi(x) = (\varphi(x)_{11}, \ldots, \varphi(x)_{1q}, \ldots, \varphi(x)_{l1}, \ldots, \varphi(x)_{lq})$. Then $\varphi : V_n \to l_1^{l_q}$ is an $O(\log n)$-distortion l_1-embedding with a probability of at least $\frac{1}{2}$. For a complete proof, see the books by Matousek [64] or Vazirani [82].

The dimension m of l_1^m in the original Bourgain proof was exponential. It has been reduced to $O(\log^2 n)$ by Linial, London and Rabinovich (1995) using Chernoff bounds.

Theorem 3.2 (Matousek, 1996). *For an integer $b > 6$, set $\beta = 2b - 1$. Then any n-point metric space (V_n, d) can be embedded into l_∞^m with distortion β, where $m = O(bn^{\frac{1}{6}} \log n)$*

Recall that (V_n, d) embeds into l_∞^n without any distortion according to Frechet's theorem. For the special case $\beta = O(\log n)$, Matousek's result implies that (V_n, d) embeds into $L_\infty^{O(\log^2 n)}$ with $O(\log n)$ distortion.

Planar metric. A finite metric space (V_n, d) is planar if (V_n, d) isometrically embeds (without any distortion) into a planar graph.

Theorem 3.3 (Rao, 1999). *Any n-point planar metric can be embedded into l_2 with distortion $O(\sqrt{\log n})$.*

Newman & Rabinovich (2002) established that the $O(\sqrt{\log n})$ bound is sharp for the so-called diamond graph (known also as Laakso's fractal).

Open Question (Linial). Is there a constant c such that any planar metric embeds into l_1 with distortion $\leq c$?

Open Question (Linial, Rabinovich). Is NP-complete or polynomial to decide if a finite metric space is planar?

Doubling dimension of a metric space (Assouad, 1983). The doubling constant of a metric space (X, d) is the smallest value λ such that every ball in X can be covered by λ balls of half the radius. The doubling dimension of (X, d) is $\log_2 \lambda$. The doubling dimension of m-dimensional l_p-space is roughly m.

Theorem 3.4 (Assouad, 1982). *If the doubling dimension of a metric space (X, d) is bounded, then for any $0 < \alpha < 1$ the metric space (X, d^α) (the snow-flaked version of d) embeds into l_2^m with distortion β, where m and β depend only on the doubling dimension of X.*

Assouad (1983) conjectured that the previous Theorem holds even when $\alpha = 1$, but Semmes (1996) disproved this conjecture. Gupta, Krauthgamer, and Lee (2003) established that the result holds for trees with doubling dimension $< \infty$.

3.3 *Johnson-Lindenstrauss flattening lemma and its relatives*

We continue with the following classical dimension-reduction theorem known as the Johnson-Lindenstrauss flattening lemma:

Theorem 3.5 (Johnson & Lindenstrauss, 1989). *Let X be an n-point set in Euclidean space, and let $\epsilon \in (0,1]$ be given. Then X can be embedded into $l_2^{O(\epsilon^{-2}\log n)}$ with distortion $(1+\epsilon)$.*

The embedding φ is obtained in the following way. Set $m = \frac{200 l_n n}{\epsilon^2}$ and let L be a random m-dimensional linear subspace of l_2^n. Then $\varphi : l_2^n \to L$ is the orthogonal projection onto L and take as the embedding the restriction of φ to X.

Theorem 3.5 can be viewed as a dimensionality reduction result: a set of points in a high dimensional space is mapped to a space with low dimension, while (approximately) preserving important characteristics of the point set.

The following result, shows (to some surprise) that no flattening lemma holds for l_1-spaces.

Theorem 3.6 (Brinkman & Charikar, 2003). *There exists an n-point subset $X \subseteq l_1$ such that for any $\beta > 1$, if X embeds into l_1^m with distortion β, then $m \geq n^{\Omega(\frac{1}{\beta^2})}$. In other words, the dimensionality reduction is impossible in l_1.*

A short proof of Theorem 3.6 was given by Lee and Naor in 2004.

3.4 *Dominating metrics*

Let d,d' be two metrics on the same set X. Then d' *dominates* d if $d'(x,y) \geq d(x,y)$ for all $x, y \in X$. Let $\mathcal{D}$ be a finite family of metrics on the same set X and let $\mathcal{P}$ be a probability distribution over $\mathcal{D}$. Then $(\mathcal{D},\mathcal{P})$, β*-probabilistically approximates* a metric d on X if

(i) every metric $d_i \in \mathcal{D}$ dominates d;
(ii) $\forall x, y \in X$, $E_{d_i \in \mathcal{D}}[d_i(x,y)] = \sum_{i=1}^{k} p_i d_i(x,y) \leq \beta d(x,y)$.

Example 3.7. Let d be the graph metric of the cycle C_n. Let d_i be the graph metric of the path obtained by removing the ith edge of C_n. Then $\frac{1}{6}\sum_{i=1}^{6} d_i \leq 3d$.

Theorem 3.8 (Fakcharoenphol, Rao, Talwar, 2003). *Let (X,d) be a finite metric space. Then there exists a set $\mathcal{D} = \{d_1, \ldots, d_r\}$ of tree metrics on X and a probability distribution $\mathcal{P}$ over $\mathcal{D}$ such that $(\mathcal{D},\mathcal{P})$, $O(\log n)$-probabilistically approximates d, where $n = |X|$. If d is a metric of a weighted graph G, then the tree metrics can be chosen to be spanning trees of G.*

Notice that Theorem 3.8 improves on previous results of Bartal (1998).

4. Graph classes defined by distance properties

In this section, we introduce and characterize main graph classes from the metric point of view related to l_1, l_2, l_∞-metrics. Namely, we will consider the following classes of graphs defined by metric properties: median graphs, Helly graphs, bridged graphs, weakly median graphs, isometric subgraphs of hypercubes and Hamming graphs, l_1-graphs, lopsided set systems and graphs, and basis graphs of matroids and Δ-matroids. The exposition follows the survey [12].

4.1 *Basic notions*

Let $G = (V, E)$ be a connected not necessarily finite, undirected and un-weighted graph endowed with the standard graph distance $d(u, v) = d_G(u, v)$. The *distance* $d(u, v) := d_G(u, v)$ between two vertices u and v is the length of a shortest (u, v)-path, and the *interval* $I(u, v)$ between u and v consists of all vertices on shortest (u, v)–paths, that is, of all vertices (metrically) *between* u and v: $I(u, v) := \{x \in V : d(u, x) + d(x, v) = d(u, v)\}$. An induced subgraph of G (or the corresponding vertex set A) is called *convex* if it includes the interval of G between any of its vertices. A *half-space* is a convex set with convex complement. An induced subgraph H of G is *isometric* if the distance between any pair of vertices in H is the same as that in G. In particular, convex subgraphs are isometric. A subgraph H of G is *gated* if for any $x \notin H$ there exists, $x' \in H$ such that $x' \in I(x, y)$ for all $y \in H$. Gated subgraphs are convex. The *ball* (or disk) $N_r(x)$ of center x and radius $r \geq 0$ consists of all vertices of G at distance at most r from x. In particular, the unit ball $N_1(x)$ comprises x and the neighborhood $N(x)$. The ball $N_r(S)$ centered at a convex set S is the union of all balls $N_r(x)$ with centers x from S.

A graph $G = (V, E)$ is *isometrically embeddable* into a graph $H = (W, F)$ if there exists a mapping $\varphi : V \to W$ such that $d_H(\varphi(u), \varphi(v)) = d_G(u, v)$ for all vertices $u, v \in V$. A *retraction* φ of H is an idempotent nonexpansive mapping of H into itself, that is, $\varphi^2 = \varphi : W \to W$ with $d(\varphi(x), \varphi(y)) \leq d(x, y)$ for all $x, y \in W$. The subgraph of H induced by the image of H under φ is referred to as a *retract* of H. Retracts are isometric subgraphs, but the converse is not true in general: the 6-cycle C_6 is an isometric subgraph but not a retract of the 3-cube H_3. Notice that in previous definitions, the host graph $G = (X, E)$ can be replaced by an arbitrary metric space (X, d). Main host spaces for isometric embedding of graphs are l_1 and l_∞-spaces, hypercubes, Hamming graphs, half-cubes, Johnson graphs, l_1 and l_∞-grids.

The *Cartesian product* $G = G_1 \square \cdots \square G_n$ has the n-tuples $(x_1, \ldots, x_n)$ as its vertices (with vertex x_i from G_i) and an edge between two vertices $x = (x_1, \ldots, x_n)$ and $y = (y_1, \ldots, y_n)$ if and only if, for some i, the vertices x_i and y_i are adjacent in G_i, and $x_j = y_j$ for the remaining $j \neq i$. Obviously, $d_G(u, v) = \sum_{i=1}^{n} d_{G_i}(u_i, v_i)$ for any two vertices $u = (u_1, \ldots, u_n)$ and $v = (v_1, \ldots, v_n)$ of G. In particular, *hypercubes* are the Cartesian powers of K_2, the complete graph on two vertices. More generally, a Cartesian product of complete graphs is a *Hamming graph.* The *half-cube* $\frac{1}{2}H_n$ is the graph whose vertex set is the collection of all subsets of X which have the same cardinality modulo 2, and two vertices A, B are adjacent in $\frac{1}{2}H_n$ exactly when $A \Delta B$ is a doubleton. The *Johnson graph* $J_{n,k}$ is the (isometric) subgraph of $\frac{1}{2}H_n$ induced by the family of all subsets of cardinality k. The *m-octahedron* (alias hyperoctahedron or Cocktail-party graph) $K_{m\times 2}$ is the complete multipartite graph with m parts, each of size 2. Thus $K_{m\times 2}$ is obtained from the complete graph K_{2m} by deleting a perfect matching.

The graphs give rise to cubical and simplicial complexes obtained by replacing every graphic cube by a unit solid cube and any clique by a solid simplex. One can derive polygonal complexes from planar graphs: for this, replace every interior face by a regular polygon with unit side. Given a graph G, denote by $|G|$ the respective cubical, simplicial, or polygonal complex derived from G. Such polyhedral complexes endowed with any of the intrinsic l_1, l_2, or l_∞-metrics constitute geodesic (isometric) subspaces of normed spaces carrying the same type of norm.

4.2 *Median graphs*

A graph G is called *median* if the interval intersection $I(x, y) \cap I(y, z) \cap I(z, x)$ is a singleton, comprising the *median* $m(x, y, z)$ for each triplet x, y, z of vertices. A systematic investigation of median graphs was initiated by Mulder [65–67]. Basic examples of median graphs are trees (which are successive point amalgams of K_2) and hypercubes (which are Cartesian powers of K_2). In fact, every median graph is obtained from hypercubes via gated amalgams. Notice also that median graphs are exactly the retracts of hypercubes and also can be characterized in truly algebraic ways.

As to their geometry, let $|G|$ is a cubical complex obtained from a median graph G. In case of median graphs, $|G|$ have many nice properties. $|G|$ equipped with the l_1-metric is a median space and therefore is an l_1-subspace. If $|G|$ carries the intrinsic l_∞-metric instead, then the resulting metric space is injective. Finally, if we impose the intrinsic l_2-metric on $|G|$, we obtain a metric space with global non-positive curvature. A *geodesic triangle* $\Delta(x_1, x_2, x_3)$ in a geodesic metric space (X, d) consists of three points in X (the vertices of Δ) and a geodesic between each pair of vertices (the edges of Δ). A *comparison triangle* for $\Delta(x_1, x_2, x_3)$ is a triangle $\Delta(x_1', x_2', x_3')$ in the Euclidean plane $\mathbb{E}^2$ such that $d_{\mathbb{E}^2}(x_i', x_j') = d(x_i, x_j)$ for $i, j \in \{1, 2, 3\}$. A geodesic metric space (X, d) is defined to be a *CAT(0) space* [25] if all geodesic triangles $\Delta(x_1, x_2, x_3)$ of X satisfy the following axiom:

If y is a point on the side of $\Delta(x_1, x_2, x_3)$ with vertices x_1 and x_2 and y' is the unique point on the line segment $[x_1', x_2']$ of the comparison triangle $\Delta(x_1', x_2', x_3')$ such that $d_{\mathbb{E}^2}(x_i', y') = d(x_i, y)$ for $i = 1, 2$, then $d(x_3, y) \le d_{\mathbb{E}^2}(x_3', y')$.

CAT(0) spaces can be characterized in several different natural ways. A geodesic metric space (X, d) is CAT(0) if and only if any two points of this space can be joined by a unique geodesic. CAT(0) is also equivalent to convexity of the function $f : [0, 1] \to X$ given by $f(t) = d(\alpha(t), \beta(t))$, for any geodesics α and β (which is further equivalent to convexity of the neighborhoods of convex sets). For more details on CAT(0) spaces see the book by Bridson and Haefliger [25]. It turns that the cubical complex $|G|$ obtained from a graph G is CAT(0) if and only if G is median.

Summarizing, here is a list of main characterizations and properties of median graphs:

Theorem 4.1 (Characterizations of median graphs).

(i) (Isbell, 1980) *Median graphs are precisely the graphs which are obtained from cubes via successive gated amalgams;*
(ii) (Bandelt, 1984) *Median graphs are precisely the retracts of hypercubes;*
(iii) (Schaefer, 1978) *Median graphs are precisely the connected components of solutions of 2-SAT instances;*
(iv) (Mulder, 1978) *Median graphs are precisely the graphs which are obtained from one vertex via successive convex expansions.*
(v) (Avann, 1961) *The median operator of a median graph satisfies the following equations:*

$$\begin{array}{ll} (M1) & (uvv) = v; \\ (M2) & (uvw) = (uwv); \\ (M3) & (uvw) = (vuw); \\ (M4) & (uv(uwx)) = (u(uvw)x). \end{array}$$

Conversely, every ternary algebra satisfying (1),(2) and (3) comes from a median graph;
(vi) (Chepoi, 2000) *G is a median graph if and only if the cubical complex $|G|$ is $CAT(0)$;*
(vii) (Gromov, 1987) *A cubical complex $|G|$ is $CAT(0)$ if and only if $|G|$ is simply connected and satisfies the following combinatorial condition: if three $(k+2)$-cubes intersect in a k-cube and pairwise intersect in $(k+1)$-cubes, then they are contained in a $(k+3)$ - cube;*
(viii) (van de Vel, 1998) *If G is median, then $(|G|, l_1)$ is an l_1-subspace;*
(ix) (Mai & Tang, 1983) *If G is median, then $(|G|, l_\infty)$ is an absolute retract, i.e., a retract of every space in which it embeds isometrically.*

4.3 *Helly graphs and hyperconvex spaces*

Hyperconvex spaces are the Menger-convex spaces in which any collection of pairwise intersecting closed balls has a point in common. It is well known that (X, d) is hyperconvex if and only if it is an absolute retract, that is, (X, d) is a retract of every metric space into which it embeds isometrically. Helly graphs are the discrete analogues of hyperconvex spaces: a graph G is thus called a *Helly graph* if the family of balls of G has the *Helly property*, that is, every collection of pairwise intersecting balls of G has a nonempty intersection. In perfect analogy with hyperconvexity, there is a close relationship between Helly graphs and absolute retracts.

A *clique-Helly graph* is a graph in which the collection of cliques (maximal simplices) has the Helly property. A vertex x of a graph G is *dominated* by another vertex y if the unit ball $N_1(y)$ includes $N_1(x)$. A finite graph G is *dismantlable* if its vertices can be linearly ordered, $v_1, v_2, \ldots, v_n$, so that, for each v_i, $i > 1$, there is a neighbor v_j, $j < i$, of v_i dominating the vertex v_i in the subgraph G_i of G induced by the vertices $v_1, \ldots, v_i$.

Theorem 4.2.

(i) (Aronszajn & Panitchpakdi, 1956) *Hyperconvex spaces are exactly the absolute retracts in the category of metric spaces or equivalently they are the retracts of* l_∞*-spaces;*
(ii) (Nowakowski & Rival, 1983) *Helly graphs are exactly the absolute retracts in the category of (reflexive) graphs or equivalently they are the retracts of the direct products of paths;*
(iii) (Bandelt & Prisner, 1991) *Helly graphs are exactly the dismantlable clique-Helly graphs.*

Examples of Helly graphs are l_∞-grids and more generally take a median graph G and replace every maximal cube with a clique, then the resulting graph is Helly.

Theorem 4.3 (Isbell 1964; Dress, 1984). *For every finite metric space* (V_n, d) *there exists the smallest hyperconvex space containing* (V_n, d) *as an isometric subspace (the tight span or the injective hull of* d*).*

4.4 *Bridged graphs*

A graph is called *bridged* if all isometric cycles of G have length three. In particular, all chordal graphs are bridged.

Theorem 4.4. *For a graph* $G = (V, E)$*, the following statements are equivalent:*

(i) *G is bridged;*
(ii) (Soltan & Chepoi, 1983; Farber & Jamison, 1987) *The balls* $N_r(S)$, $r \geq 1$, *centered at convex sets S are convex;*
(iii) (Chepoi, 2000) *The simplicial complex* $|G|$ *is simply connected and for every vertex* v, $N_1(v)$ *does not contain induced 4-cycles and 5-cycles;*
(iv) (Anstee & Farber, 1988) *G is precisely a dismantlable graph without induced 4- and 5-cycles;* (Chepoi, 1997) *The dismantling scheme is provided by BFS.*

Examples of bridged graphs are chordal graphs, graphs for which the simplicial complex $|G|$ is 2-dimensional and $CAT(0)$, and planar graphs in which all inner faces are triangles and all inner vertices have degree ≥ 6.

4.5 *Weakly median graphs*

A graph G is called *weakly modular* if it satisfies the following conditions:

(*triangle condition*) for any three vertices u, v, w with $1 = d(v, w) < d(u, v) = d(u, w)$ there exists a common neighbor x of v and w such that $d(u, x) = d(u, v) - 1$;

(*quadrangle condition*) for any four vertices u, v, w, z with $d(v, z) = d(w, z) = 1$ and $2 = d(v, w) \leq d(u, v) = d(u, w) = d(u, z) - 1$, there exists a common neighbor x of v and w such that $d(u, x) = d(u, v) - 1$.

The graph G is *weakly median* if G is weakly modular and for every three or four vertices occurring in the triangle condition or quadrangle condition, the vertex x is unique.

Three vertices v_1, v_2, and v_3 form a *metric triangle* $v_1v_2v_3$ if the intervals $I(v_1, v_2)$, $I(v_2, v_3)$ and $I(v_3, v_1)$ pairwise intersect only in the common end vertices. If $d(v_1, v_2) = d(v_2, v_3) = d(v_3, v_1) = k$, then this metric triangle is called *equilateral* of *size* k. Then a graph G is weakly modular if and only if for every metric triangle uvw all vertices of the interval $I(v, w)$ are at the same distance $k = d_G(u, v)$ from u. A metric triangle uvw of G is a *quasi–median* of the triplet x, y, z if the following metric equalities are satisfied:

$$d(x, y) = d(x, u) + d(u, v) + d(v, y),$$
$$d(y, z) = d(y, v) + d(v, w) + d(w, z),$$
$$d(z, x) = d(z, w) + d(w, u) + d(u, x).$$

Every triplet x, y, z of a graph has at least one quasi-median. One can show that a weakly modular graph is weakly median if and only if for each triplet of vertices there exists a unique quasi-median.

Theorem 4.5 (Bandelt & Chepoi, 2000, 2006). *Every weakly median graph can be obtained from the following prime graphs using Cartesian products and gated amalgams: K_2, the 5-wheel, induced subgraphs of hyperoctahedra that include either K_4 or a 4-wheel, and two-connected K_4- and $K_{1,1,3}$-free bridged graphs are the prime finite weakly median graphs. The latter bridged graphs are exactly the graphs which can be realized as plane graphs such that all inner faces are triangles and all inner vertices have degrees larger than 5.*

Proposition 4.6 (Bandelt & Chepoi, 2000). *Every weakly median graph G is l_1-embeddable. G has a scale 2 embedding in a hypercube if and only if it does not contain an induced K_6 minus an edge.*

The bridged prime constituents of weakly median graphs are therefore exactly the (3,6)-graphs (see the next section) that have only triangles as their inner faces.

In order to investigate weakly median graphs from an algebraic point of view, notice that every graph G with vertex set V can be turned into a ternary algebra, called an *apex algebra* of G [19]: an *apex operation* $(\dots) : V^3 \to V$ maps any triplets x, y, z and x, z, y to some vertex $u = (xyz) = (xzy) \in I(x, y) \cap I(x, z)$ such that $I(x, u)$ is maximal with respect to inclusion. For weakly median graphs the apex algebra is uniquely defined, because for any triplet x, y, z the vertices (xyz), (yxz), and (zxy) form the unique quasi-median of the triplet. The equational description follows the lines of median algebras:

Theorem 4.7 (Bandelt & Chepoi, 2006). *Weakly median graphs with their apex operations are exactly the discrete ternary algebras satisfying the four axioms (M1), and*

(M5) $(uv(uwx)) = (ux(uvw))$
(M6) $((uwx)(vwx)u) = (uwx)$,
(M7) $((wux)(uvw)v) = (w(uvw)(vu(wux)))$.

4.6 l_1-graphs

In this subsection we present an overview of classes of graphs which can be isometrically embedded into hypercubes, Hamming graphs, and in l_1-spaces. We start with canonical embedding of a graph into a Cartesian product. Any graph may be isometrically embedded into different Cartesian products, however, there is a finest choice, viz., the *canonical isometric embedding* [45]. Two edges xy and uv of a graph $G = (V, E)$ are in the Djokovic-Winkler relation Θ if $d(x, u) + d(y, v) \neq d(x, v) + d(y, u)$. This relation on the edge set E is reflexive and symmetric but not necessarily transitive. Let Θ^* denote the transitive closure of Θ, and let $E_1, \dots, E_k$ be the blocks of Θ^*. Let G_i $(i = 1, \dots, k)$ be the graph having the connected components of the graph $(V, E - E_i)$ as its vertices, with two different components being adjacent when connected by an edge from E_i; alternatively, one can view G_i as the graph resulting from the contraction of all edges in $E - E_i$. This contraction induces a natural projection α_i from G onto G_i.

Theorem 4.8 (Graham & Winkler, 1985). *The map $\alpha : G \to G_1 \square \dots \square G_k$ defined by $\alpha(v) = (\alpha_1(v), \dots, \alpha_k(v))$ constitutes an isometric embedding, which is the finest isometric embedding of G into a Cartesian product (whence the name "canonical").*

Every edge ab of a graph $G = (V, E)$ induces a partition of V into $V = a/b \cup a\backslash b \cup a|b$, where $a/b = \{v \in V : d(v, a) < d(v, b)\}$ and $a|b = \{v \in V : d(v, a) = d(v, b)\}$. Clearly, if the graph G is bipartite, then $a|b = \emptyset$ for each $ab \in E$.

Theorem 4.9 (Djokovic, 1973). *For a bipartite graph G the following conditions are equivalent:*

(i) *G can be isometrically embedded into a hypercube;*
(ii) *G is an l_1-graph;*
(iii) *$a/b, a\backslash b$ are convex for each edge ab of G;*
(iv) *the relation Θ is transitive on the edge set of G.*

Theorem 4.10. *Let G be a graph.*

(i) (Chepoi, 1988) *G can be isometrically embedded into a Hamming graph if and only if the sets a/b, $a\backslash b$, $a/b \cup a|b$ and $a\backslash b \cup a|b$ are convex for all edges ab of G;*
(ii) (Winkler, 1984) *G can be isometrically embedded into a Hamming graph $H_{3,3,\dots,3}$ (Cartesian power of K_3) if and only if the relation Θ is transitive.*

The dimension k of the smallest hypercube into which a finite graph $G = (V, E)$ may be isometrically embedded equals the number of blocks $E_1, \dots, E_k$ of the equivalence relation Θ. For other characterizations of isometric subgraphs of hypercubes see [38,48]. Recognition algorithms for graphs that are isometrically embeddable into (hypercubes and) Hamming graphs are surveyed in [48].

A graph $G = (V, E)$ is *scale λ embeddable* into a graph $H = (W, F)$ for some positive integer λ if there exists a mapping $\varphi : V \to W$ such that $d_H(\varphi(u), \varphi(v)) = \lambda d_G(u, v)$ for all vertices $u, v \in V$ [38]. In the particular case $\lambda = 1$ we obtain the notion of isometric embedding. Hamming graphs as well as half-cubes and the Johnson graphs are scale 2 embedded in hypercubes. On the other hand, the m-octahedron $K_{m\times 2}$ is scale embeddable into a hypercube, but its scale grows with m.

Proposition 4.11 (Assouad & Deza, 1980). *G is an l_1-graph if and only if it admits a scale embedding into a hypercube.*

This result paves the way to a description of l_1-graphs via the canonical embedding.

Theorem 4.12. *Let G be a graph.*

(i) (Shpectorov, 1993) *G is an l_1-graph if and only if G is an isometric subgraph of a Cartesian product of half-cubes and hyperoctahedra;*
(ii) (Chepoi, Deza, Grishukhin, 1997) *Every K_5-free l_1-graph is isometrically embeddable into a half-cube, and every K_4-free l_1-graph is l_1-rigid. In particular, every planar l_1-graph is isometrically embeddable into a half-cube.*

Some particular classes of plane graphs which can be isometrically embedded into a half-cube are worth mentioning explicitly:

(4,4)-graphs [69] *are plane graphs in which all inner faces have lengths* ≥ 4 *and all inner vertices have degrees* ≥ 4;

(6,3)-graphs [36] *are plane graphs in which all inner faces have length* ≥ 6 *and all inner vertices have degrees* ≥ 3;

(3,6)-graphs [36] *are plane graphs in which (all inner faces have lengths* ≥ 3 *and) all inner vertices have degrees* ≥ 6.

The (4,4)-, (3,6)-, and (6,3)-graphs have been introduced in the context of combinatorial group theory and all of them generate CAT(0) polygonal complexes.

Isometric subgraphs of hypercubes naturally arise in several contexts. An (affine) *arrangement of hyperplanes* in $\mathbb{R}^k$ is a finite family $\mathcal{A} = (H_i)_{i\in I}$ of (affine) hyperplanes in $\mathbb{R}^k$. One can choose a "positive side" for every hyperplane in $\mathcal{A}$, which is realized by a sign vector $s(x) \in \{+, -, 0\}^I$ for every point $x \in \mathbb{R}^n$, where $x_i(e)$ denotes whether x is on the positive side of H_i, on the negative side, or lies on H_i. The set of all points $x \in \mathbb{R}^k$ having the same sign vector s forms a *cell* in the decomposition of $\mathbb{R}^k$ induced by $\mathcal{A}$. The maximal (i.e., the n-dimensional) cells are called *regions* (or *topes*). The *tope graph* (or *graph of regions*) $\mathcal{T}(\mathcal{A})$ of an arrangement $\mathcal{A}$ has as vertices the set of regions of $\mathcal{A}$ and as edges the pairs of regions which share a common facet. Tope graphs of any arrangements of n hyperplanes embed isometrically into the hypercube H_n, namely, the graph distance between two regions equals the number of hyperplanes of $\mathcal{A}$ which separate these regions.

Graphs for which the shortest-path metrics are totally decomposable constitute certain low-dimensional polyhedral complexes. In the finite bipartite case, they are known to be built up from cycles (their "cells"), whence the name "cellular".

Theorem 4.13 (Bandelt & Chepoi, 1996). *For a bipartite graph $G = (V, E)$ with at least two vertices, the following conditions are equivalent:*

(i) *G is cellular;*

(ii) *every isometric cycle of G is gated, and G does not contain any three isometric cycles C_1, C_2, C_3 and three distinct edges e_1, e_2, e_3 sharing a common vertex such that e_i belongs to C_j exactly when $i \neq j$;*
(iii) *G can be obtained from a collection of single edges and even cycles by successive gated amalgamations;*
(iv) *the system of splits $\{u/v, u\backslash v\}$ is triplewise weakly compatible.*

In the context of chemical graph theory, a *benzenoid* is a particular (6,3)-graph, viz., a plane graph in which all inner faces are regular hexagons and all inner vertices have degree 3. These graphs are not necessarily cellular but they admit isometric embeddings into hypercubes [58] and, in particular, into the Cartesian product of three trees. With the latter coordinatization in hand, one can compute the Wiener index (the sum of pairwise distances) of benzenoids in optimal time [48].

4.7 *Lopsided subsets of hypercubes*

The lopsided sets have been introduced by J. Lawrence [59] in order to investigate the subgraphs $\mathcal{S}$ of the n-cube with vertex set $\{0, 1\}^n$ which encode the intersection pattern of a given convex set K in n-dimensional Euclidean space with the (closed) orthants of $\mathbb{R}^n$. Specifically, a vertex x of H_n belongs to $\mathcal{S} = \mathcal{S}(K)$ exactly when K meets the orthant of $\mathbb{R}^n$ in which the ± 1 ("sign") vector $2x - \mathbf{1}$ lies. Every subgraph $\mathcal{S}$ of H_n which arises in this way is hereditarily asymmetric in the following sense (whence the name "lopsided") [59]:

(Lawrence condition) *each face F of H_n which intersects $\mathcal{S}$ properly includes a vertex pair u, v diametrical in F such that $\mathcal{S}$ contains exactly one of u and v.*

One can then define $\mathcal{S}$ to be *lopsided* if it satisfies the preceding characteristic property. Lopsided sets induce isometric subgraphs of H_n. Moreover, a hereditary version of isometry ("superisometry" [13]) then characterizes lopsidedness. To express this, factor the n-cube 2^X hosting $\mathcal{S}$ into 2^Y and 2^{X-Y} with respect to any subset Y of $X = \{1, \ldots, n\}$ and let the subset $\mathcal{S}^Y$ of 2^{X-Y} encode the location of all 2^Y fibers within $\mathcal{S}$:

$$\mathcal{S}^Y := \{t \in 2^{X-Y} : \text{ every extension } s \in 2^X \text{ of } t \text{ belongs to } \mathcal{S}\}.$$

The coordinate subsets Y of X for which some 2^Y fiber exists in $\mathcal{S}$ form a simplicial complex:

$$\underline{\mathcal{X}}(\mathcal{S}) := \{Y \subseteq X : \mathcal{S}^Y \text{ is nonempty }\}.$$

The sets $\mathcal{S}^Y$ and $\underline{\mathcal{X}}(\mathcal{S})$ can be compared to the sets

$$\mathcal{S}_Y := \mathcal{S}|_{X-Y} \text{ and } \overline{\mathcal{X}}(\mathcal{S}) := \{Y \subseteq X : \mathcal{S}_Y = 2^{X-Y}\},$$

which determines the image of $\mathcal{S}$ under the projection onto 2^{X-Y} and collects the coordinate subsets Y such that the projections onto 2^{X-Y} are surjective, respectively.

The original definition of a lopsided set in [59] can be expressed in terms of the simplicial complexes $\overline{\mathcal{X}}(\mathcal{S})$ and $\underline{\mathcal{X}}(\mathcal{S})$ as the requirement that for each $Y \subseteq X$ the alternative $Y \in \underline{\mathcal{X}}(\mathcal{S})$ or $Y \notin \overline{\mathcal{X}}(\mathcal{S})$ holds. One of the basic observations is that the inequality $\#\underline{\mathcal{X}}(\mathcal{S}) \leq \#\mathcal{S} \leq \#\overline{\mathcal{X}}(\mathcal{S})$ holds for all subsets $\mathcal{S}$ of 2^X [13]. The following theorem of [13] presents the basic combinatorial and graph-theoretical characterizations of lopsided sets.

Theorem 4.14 (Bandelt, Chepoi, Dress, Koolen, 2006). *For a subset $\mathcal{S}$ of 2^X the following conditions are equivalent:*

(i) *$\mathcal{S}$ is lopsided;*
(ii) *$2^X - \mathcal{S}$ is lopsided;*
(iii) *$\mathcal{S}^Y$ is isometric for all $Y \subseteq X$;*
(iv) $\#\mathcal{S} = \#\overline{\mathcal{X}}(\mathcal{S})$;
(v) $\#\mathcal{S} = \#\underline{\mathcal{X}}(\mathcal{S})$;
(vi) $\underline{\mathcal{X}}(\mathcal{S}) = \overline{\mathcal{X}}(\mathcal{S})$;
(vii) *$\mathcal{S}$ is connected, and $\mathcal{S}^{\{e\}}$ is lopsided for every $e \in X$;*
(viii) *$\mathcal{S}$ is isometric, and both $\mathcal{S}_{\{e\}}$ and $\mathcal{S}^{\{e\}}$ are lopsided for some $e \in X$.*

As to the geometric interpretation of lopsided sets, it was already noted in [59] that not every lopsided set encodes the orthant intersection pattern for a convex set in Euclidean space. In order to have a full geometric representation, one has to resort to a weaker concept ("ortho-convexity") of convexity. For a subset $\mathcal{S}$ of 2^X, let $|\mathcal{S}|$ be the polyhedral cubical complex obtained by replacing all faces of $\mathcal{S}$ by solid cubes. If $\mathcal{S}$ is connected, then $|\mathcal{S}|$ is connected as well, and therefore can be endowed with an intrinsic l_1-metric $d_{|\mathcal{S}|}$. The resulting metric space $(|\mathcal{S}|, d_{|\mathcal{S}|})$ is geodesic but not necessarily a metric subspace of $(\mathbb{R}^X, \|\cdot\|_1)$. In fact, l_1-isometry of $|\mathcal{S}|$ is yet another characteristic feature of lopsidedness of $\mathcal{S}$, as the following result of [16] shows:

Theorem 4.15 (Bandelt, Chepoi, Dress, Koolen, 2006). *For a subset $\mathcal{S}$ of 2^X the following conditions are equivalent:*

(i) *$\mathcal{S}$ is lopsided;*
(ii) *$|\mathcal{S}|$ endowed with the intrinsic l_1–metric $d_{|\mathcal{S}|}$ is a metric subspace of $(\mathbb{R}^X, \|\cdot\|_1)$;*
(iii) *$\mathcal{S}$ is isometric in 2^X and every face of $|\mathcal{S}|$ is a gated subset of $(|\mathcal{S}|, d_{|\mathcal{S}|})$;*
(iv) *$\mathcal{S}$ encodes the orthant intersection pattern for some geodesic metric subspace K of $(\mathbb{R}^n, \|\cdot\|_1)$.*

Examples of lopsided sets:

(i) set systems obtained from median graphs;
(ii) tope graphs of arrangements of pseudolines such that no more than two pseudolines meet at any intersection point;
(iii) antimatroids, also known under the name convex geometries;
(iv) signed maps of regions of simple affine arrangements of hyperplanes;
(v) vertex sets of rhombic tilings;
(vi) maximum ("circular") set systems yielding weakly compatible split systems;
(vii) maximum set systems of a given Vapnik-Chervonenskis dimension.

4.8 *Basis graphs of matroids and even Δ-matroids*

A *matroid* of rank k on a set I of n elements is a collection $\mathcal{B}$ of subsets of I of size k, called *bases,* which satisfy the following *exchange property: for* $A, B \in \mathcal{B}$ *and* $i \in A - B$ *there exists* $j \in B - A$ *such that* $(A - \{i\}) \cup \{j\} \in \mathcal{B}$. One says that the base $A - \{i\} \cup \{j\}$ is obtained from the base A by an *elementary exchange*. The *basis graph* $G = G(\mathcal{B})$ of a matroid $\mathcal{B}$ is the graph whose vertices are the bases of $\mathcal{B}$ and edges are the pairs A, B of bases differing by an elementary exchange. The exchange property implies that $G(\mathcal{B})$ is an isometric subgraph of the Johnson graph $J_{n,k}$.

A Δ-*matroid* [23,29,39] is a collection $\mathcal{B}$ of subsets of a set I with $\#I = n$, called *bases*, not necessarily equi-cardinal, satisfying the following *symmetric exchange property: for* $A, B \in \mathcal{B}$ *and* $i \in A\Delta B$, *there exists* $j \in B\Delta A$ *such that* $A\Delta\{i, j\} \in \mathcal{B}$. A Δ-matroid whose bases all have the same cardinality modulo 2 is called an *even* Δ-*matroid*. The matroids are precisely the (even) Δ-matroids for which all members of $\mathcal{B}$ have the same cardinality. For a subset J of I denote $\mathcal{B}\Delta J := \{B\Delta J : B \in \mathcal{B}\}$. If A, B are two bases of an even Δ-matroid $\mathcal{B}$ and $B = A\Delta\{i, j\}$ we say that B is obtained from A by an *elementary exchange*. The *basis graph* $G = G(\mathcal{B})$ of an even Δ-matroid $\mathcal{B}$ is the graph whose vertices are the bases of $\mathcal{B}$ and edges are the pairs A, B of bases such that $\#A\Delta B = 2$. It is clear that $G(\mathcal{B})$ is an isometric subgraph of the half-cube $\frac{1}{2}H_n$.

The following two conditions play a key role in the characterizations of basis graphs. A graph G satisfies the *interval condition* (ICm) ($m \geq 2$) if for any vertices u, v at distance 2, the interval $I(u, v)$ includes an induced square $C_4 = K_{2\times 2}$ and is contained in the m-octahedron $K_{m\times 2}$. A graph G satisfies the *positioning condition* (PC) if for each vertex b and each square $v_1v_2v_3v_4$ of G the equality $d(b, v_1) + d(b, v_3) = d(b, v_2) + d(b, v_4)$ holds. Obviously, a graph in which every interval between vertices at distance 2 includes a 4-cycle satisfies the positioning condition exactly when G is meshed.

Theorem 4.16. *Let G be a finite graph.*

(i) (Maurer, 1973) *G is a basis graph of a matroid if and only if it satisfies the interval condition* (IC3), *the positioning condition* (PC), *and the subgraph induced by the neighborhood* $N(v)$ *of any vertex* v *is the line graph of a bipartite graph;*

(ii) (Chepoi, 2006) *G is a basis graph of an even* Δ*-matroid if and only if it satisfies the interval condition* (IC4), *the positioning condition* (PC), *and the subgraph induced by the neighborhood* $N(v)$ *of any vertex* v *is the line graph of some graph.*

References

[1] R. Alexander, Planes for which the lines are the shortest paths between points, *Illinois J. Math.*, **22** (1978) 177–190.

[2] N. Alon and P. Pudlak, Equilateral sets in l_p, *Geometric and Functional Analysis*, **13** (2003) 467–482.

[3] R. V. Ambartzumian, Combinatorial Integral Geometry, John Wiley and Sons (1982).

[4] R. P. Anstee and M. Farber, On bridged graphs and cop-win graphs, *J. Combin. Th. Ser.*, **B44** (1988) 22–28.

[5] N. Aronszajn and P. Panitchpakdi, Extensions of uniformly continuous transformations and hyper convex metric spaces, *Pacific. J. Math.*, **6** (1956) 405–439.

[6] P. Assouad, Embeddability of regular polytopes and honeycombs in hypercubes, in: The geometric vein: The Coxeter Festschrift (1981) 141–147.

[7] P. Assouad and M. Deza, Espaces m'etriques plongeable dans hypercube: aspects combinatoires, *Annals of Discrete Math.*, **8** (1980) 197–210.

[8] H.-J. Bandelt, Retracts of hypercubes, *J. Graph Th.*, **8** (1984) 501–510.

[9] H.-J. Bandelt and V. Chepoi, Embedding metric spaces in the rectilinear plane: a six-point criterion, *Discrete and Computational Geometry*, **15** (1996) 107–117.

[10] H.-J. Bandelt and V. Chepoi, Cellular bipartite graphs, *Europ. J. Combin.*, **17** (1996) 121–134.

[11] H.-J. Bandelt and V. Chepoi, Decomposition and l_1-embedding of weakly median graphs, *Europ. J. Combin.*, **21** (2000) 701–714.

[12] H.-J. Bandelt and V. Chepoi, Metric graph theory and geometry: a survey, Contemporary Mathematics (to appear).

[13] H.-J. Bandelt, V. Chepoi, A. Dress and J. Koolen, Combinatorics of lopsided sets, *Europ. J. Combinatorics*, **27** (2006) 669–689.

[14] H.-J. Bandelt and V. Chepoi, The algebra of metric betweenness I: subdirect representation and retraction, *Europ. J. Combin.* (to appear).

[15] H.-J. Bandelt and V. Chepoi, The algebra of metric betweenness II: equational characterization of weakly median graphs (submitted).

[16] H.-J. Bandelt, V. Chepoi, A. Dress and J. Koolen, Geometry of lopsided sets (in preparation).

[17] H.-J. Bandelt, V. Chepoi and M. Laurent, Embedding into rectilinear spaces, *Discrete and Computational Geometry*, **19** (1998) no. 4, 595–604.

[18] H-J. Bandelt and A. W. M. Dress, A canonical decomposition theory for metrics on a finite set, *Advances in Mathematics*, **92** (1992) 47–105.

[19] H.-J. Bandelt, H. M. Mulder and E. Wilkeit, Quasi-median graphs and algebras, *J. Graph Th.*, **18** (1994) 681–703.

[20] H.-J. Bandelt and E. Prisner, Clique graphs and Helly graphs, *J. Combin. Th. Ser. B*, **51** (1991) 34–45.

[21] Y. Bartal, On approximating arbitrary metrics by tree metrics, Proceedings of the 30th Annual ACM Symposium on Theory of Computing (1998).

[22] A. Björner, M. Las Vergnas, B. Sturmfels, N. White and G. Ziegler, Oriented Matroids, Encyclopedia of Mathematics and its Applications Vol. 46, Cambridge University Press, Cambridge (1993).

[23] A. Bouchet, Greedy algorithm and symmetric matroids, *Math. Programming*, **38** (1987) 147–159.

[24] J. Bourgain, On Lipschitz embeddings of finite metric spaces in Hilbert space, *Israel Journal of Mathematics*, **52** (1985) 46–52.

[25] M. Bridson and A. Haefliger, Metric Spaces of Non-Positive Curvature, Springer, Berlin (1999).

[26] B. Brinkman and M. Charikar, On the impossibility of dimension reduction in l_1, In: Proc. 35th Annu. ACM Sympos. Theory Comput. (2003).

[27] P. Buneman, The recovery of trees from measures of dissimilarity. In: Mathematics in Archeological and Historical Sciences, (eds.) F. R. Hodson, D. G. Kendall and P. Tautu, Edinburgh University Press, Edinburgh (1971) 387–395.

[28] P. Buneman, A note on metric properties on trees, *J. Combin. Th. Ser. B*, **7** (1974) 48–50.

[29] R. Chandrasekaran and S. N. Kabadi, Pseudomatroids, *Discrete Math.*, **71** (1988) 205–217.

[30] V. Chepoi, Isometric subgraphs of Hamming graphs and d-convexity, *Cybernetics*, **24** (1988) 6–10 (Russian, English transl.).

[31] V. Chepoi, Bridged graphs are cop-win graphs: an algorithmic proof, *J. Combin. Th. Ser. B*, **69** (1997) 97–100.

[32] V. Chepoi, Separation of two convex sets in convexity structures, *J. Geometry*, **50** (1994) 30–51.

[33] V. Chepoi, Graphs of some CAT(0) complexes, *Advances Applied Mathematics*, **24** (2000) 125–179.

[34] V. Chepoi, Basis graphs of even delta-matroids, *J. Combin. Th. Ser. B*, **97** (2007) 175–192.

[35] V. Chepoi, M. Deza and V. Grishukhin, Clin dóeil on L_1-embeddable planar graphs, *Disc. Appl. Math.*, **80** (1997) 3–19.

[36] V Chepoi, F. Dragan and Y. Vaxes, Distance and routing labeling schemes for non-positively curved plane graphs, *J. Algorithms*, **61** (2007) 60–88.

[37] V. Chepoi and B. Fichet, A note on circular decomposable metrics, *Geometriae Dedicata*, **69** (1998) 237–240.

[38] M. Deza and M. Laurent, Geometry of Cuts and Metrics, Springer, Berlin (1991).

[39] A. W. M. Dress and T. Havel, Some combinatorial properties of discriminants in metric vector spaces, *Adv. Math.*, **62** (1986) 285–312.

[40] A. W. M. Dress, Trees, tight extensions of metric spaces, and the cohomological dimension of certain groups, *Adv. Math.*, **53** (1984) 321–402.

[41] D. Z. Djokovic, Distance preserving subgraphs of hypercubes, *J. Combin. Th. Ser. B*, **14** (1973) 263–267.

[42] J. Edmonds, Embedding into l_∞^2 is ieasy and into l_∞^3 is NP-hard, Discrete Computational Geometry (to appear).

[43] J. Fakcharoenphol, S. Rao and K. Talwar, A tight bound on approximating arbitrary metrics by tree metrics, Proceedings of the 35th Annual ACM Symposium on Theory of Computing (2003).

[44] B. Fichet, The role played by L_1 in data analysis. In: Statistical data analysis based on the L_1-norm and related methods, (ed.) Y. Dodge, pages 185–193. Elsevier Science Publishers B.V. (North Holland) (1987).

[45] R. L. Graham and P. M. Winkler, On isometric embeddings of graphs, *Trans. Amer. Math. Soc.*, **288** (1985) 527–536.

[46] M. Gromov, Hyperbolic groups, in Essays in Group theory, (ed.) S. M. Gersten, pp. 75–263, MSRI Publ., Vol. 8, Springer, Berlin (1987).

[47] A. Gupta, R. Krauthgamer and J. R. Lee, Bounded geometries, fractals, and low-distortion embeddings, In: Proc. of 44nd IEEE Symposium of Foundations of Computer Science (2003).

[48] W. Imrich and S. Klavžar, Product Graphs: Structure and Recognition, Wiley-Interscience Publication, New York (2000).

[49] P. Indyk, Algorithmic applications of low-distortion embeddings. In: Proc. of 42nd IEEE Symposium of Foundations of Computer Science (2001).

[50] P. Indyk and J. Matousek, Low-distortion embeddings of finite metric spaces.

[51] J. R. Isbell, Median algebra, *Trans. Amer. Math. Soc.*, **260** (1980) 319–362.

[52] W. B. Johnson and J. Lindenstrauss, Extensions of Lipschitz mappings into Hilbert space, *Contemporary Mathematics*, **26** (1984) 189–206.

[53] A. V. Karzanov, Metrics and undirected cuts, *Math. Programm.*, **32** (1985) 183–198.

[54] A. V. Karzanov, Minimum 0-extensions of graph metrics, *Europ. J. Combin.*, **19** (1998) 71–101.

[55] A. V. Karzanov, Metrics with finite sets of primitive extensions, *Ann. Combin.*, **19** (1998) 213–243.

[56] J. B. Kelly, Metric inequalities and symmetric differences. In: Inequalities II, (ed.) O. Shisha, pages 193–212, Academic Press, New York (1970).

[57] J. B. Kelly, Hypermetric spaces, In: The Geometry of Metric and Linear spaces, *Lecture Notes in Math.*, **490**, Springer-Verlag (1975) pp. 17–31.

[58] S. Klavžar, I. Gutman and B. Mohar, Labeling of benzenoid systems which reflects the vertex-distance relations, *J. Chem. Inf. Comput. Sci.*, **35** (1995) 590–593.

[59] J. Lawrence, Lopsided sets and orthant-intersection of convex sets, *Pacific J. Math.*, **104** (1983) 155–173.

[60] J. R. Lee and A. Naor, Embedding the diamond graph in L_p and dimension reduction in L_1, Manuscript, June (2003).

[61] N. Linial, E. London and Y. Rabinovich, The geometry of graphs and some of its algorithmic applications, *Combinatorica*, **15** (1995) 215–245.

[62] J. H. Mai and Y. Tang, An injective metrization for collapsible polyhedra, *Proc. Amer. Math. Soc.*, **88** (1983) 333–337.

[63] S. M. Malitz and J. I. Malitz, A bounded compactness theorem for L_1-embeddability of metric spaces in the plane, *Discrete Comput. Geom.*, **8** (1992) 373–385.

[64] J. Matousek, Lectures on Discrete Geometry, Springer Verlag, Prague (2001).
bibitemmau-73 S. B. Maurer, Matroid basis graphs I, *J. Combin. Th. Ser. B*, **14** (1973) 216–240; II, *J. Combin. Th. Ser. B*, **15** (1973) 121–145.

[65] H. M. Mulder, The Interval Function of a Graph, *Math. Centre Tracts*, **132** (Amsterdam) (1980).

[66] H. M. Mulder, The structure of median graphs, *Discrete Math.*, **24** (1978) 197–204.

[67] H. M. Mulder and A. Schrijver, Median graphs and Helly hypergraphs, *Discr. Math.*, **25** (1979) 41–50.

[68] I. Newman and Yu. Rabinovich, A lower bound on the distortion of embedding planar metrics into Euclidean space, *Discrete Comput. Geom.*, **29** (2003) 77–81.

[69] Ch. Prisăcaru, P. Soltan and V. Chepoi, On embeddings of planar graphs into hypercubes, *Proc. Moldavian Academy of Sci., Math.*, **1** (1990) 43–50 (in Russian).

[70] S. Rao. Small distortion and volume respecting embeddings for planar and Euclidean metrics, In: Proc. 15th Annu. ACM Sympos. Comput. Geom., pp. 300–306 (1999).

[71] T. J. Schaefer, The complexity of satisfability problems, Proc. 10th ACM Symp. on Theory of Computing (1978) 216–226.

[72] I. J. Schoenberg, Remarks to Maurice Frechet's article: Sur la definition axiomatique d'une classe d'espaces vectoriels distancies applicables vectoriellement sur l'espace de Hilbert, *Annals Mathematics*, **36** (1935) 724–732.

[73] I. J. Schoenberg. Metric spaces and positive definite functions, *Trans. Amer. Math. Soc.*, **44** (1938) 522–553.

[74] S. Semmes, On the nonexistence of bi-Lipschitz parameterizations and geometric problems about A-weights, *Rev. Mat. Iberoamericana*, **12** (1996) 337–410.

[75] S. V. Shpectorov, On scale embeddings of graphs into hypercubes, *Europ. J. Combin.*, **14** (1993) 117–130.

[76] J. M. S. Simoes Pereira. A note on the tree realizability of a distance matrix, *J. Combin. Theory*, **6** (1969) 303–310.

[77] V. P. Soltan and V. D. Chepoi, Conditions for invariance of set diameters under d-convexification in a graph, *Cybernetics*, **19** (1983) 750–756 (Russian, English transl.)

[78] K. A. Zaretsky, Reconstruction of a tree from the distances between its pendant vertices, *Uspekhi Math. Nauk (Russian Mathematical Surveys)* **20** (1965) 90–92 (in Russian).

[79] M. van de Vel, Matching binary convexities, *Topology Appl.*, **16** (1983) 207–235.

[80] M. van de Vel, Theory of Convex Structures, Elsevier Science Publishers, Amsterdam (1993).

[81] M. van de Vel, Collapsible polyhedra and median spaces, *Proc. Amer. Math. Soc.*, **126** (1998) 2811–2818.

[82] V. Vazirani, Approximation Algorithms, Springer, Berlin (2001).

[83] P. M. Winkler, Isometric embedding in the product of complete graphs, *Discr. Appl. Math.*, **7** (1984) 221–225.

Joint Proc. Int. Wrk.shps. Conv. in Discr. Struct. and Metric & Conv. Graph Th. No. 5, 2008, pp. 47–55.

Characterizations of J-Monotone Graphs*

Manoj Changat[1] and Joseph Mathews[2]

[1]Department of Futures Studies, University of Kerala, Thiruvananthapuram 695 034, India
[2]Department of Mathematics, S.B. College, Changanassery 686 101, India
e-mail: mchangat@gmail.com; jose_chingam@yahoo.co.in

Abstract. The Induced path transit function $J(u, v)$ in a graph consists of the set of all vertices lying on any induced path between the vertices u and v. A transit function R satisfies the *monotone axiom* if $x, y \in R(u, v)$ implies $R(x, y) \subseteq R(u, v)$ and betweenness axiom $b2$ if $x \in R(u, v)$ implies $R(u, x) \subseteq R(u, v)$. A graph G is called J-monotone if $J(u, v)$ is J-convex for every $u, v \in V(G)$. We survey the results related to the monotone axiom of the function J and give a new characterization of J-monotone graphs using forbidden induced subgraphs. Along the way, we will characterize the graphs for which the function satisfies the betweenness axiom $b2$.

Keywords. Transit function, induced path, monotone axiom, Peano axiom, proper vertex subpath, J-monotone-graphs.

Subject Classification: 05C99, 05C12, 05C38, 05C75

1. Introduction

The geodesic interval function I, the induced path function J and the corresponding convexities of a connected graph have been studied extensively by various authors ([5,7–9,11,12,14,17,18,20,21,23,24]) and constitute a developing area of metric graph theory and related areas. These functions and the corresponding convexities can be studied in a general frame work using transit functions, also known as interval functions, see for example, [1,25]. A number of prototype problems generalizing the notion of intervals and convexity in the case of graphs is being surveyed in [16]. In this paper we follow the terminology in [16].

The monotone axiom is interesting because it is a strong axiom satisfied by the usual interval in a real vector space. The notion of *interval monotone* (*I-monotone*; I stands for the geodesic interval function of G) graph was introduced by H. M. Mulder in [15]. It can be easily seen that not every graph is I-monotone, for example the complete bipartite graph $K_{2,3}$ is not I-monotone. Mulder proved that if G contains no subgraph *homomorphic* to $K_{2,3}$ or $W_5 - x$, then G is I-monotone. Mulder conjectured that the interval regular graphs ([15]) are I-monotone, but Mollard and Nomura ([17,21]) disproved this. A characterization of I-monotone graphs using forbidden induced subgraphs still remains as a challenging open problem.

The induced path transit function J of a connected graph G also need not always satisfy the monotone axiom. A connected graph G is called J-monotone if $x, y \in$

*Supported by the Department of Science and Technology, Govt. of India under the grant DST/INT/SLOV/P-03/05.

$J(u, v)$ implies $J(x, y) \subseteq J(u, v)$, for any two vertices u and v in G. Examples of graphs non-J-monotone are $K_{2,3}$ and subdivided $K_{2,3}$. The monotone axiom is also studied for the induced path transit function J and some characterizations are available, see [3,4]. The monotone and *Peano* axioms are equivalent for the induced path transit function J. A characterization of planar J-monotone graphs using the Peano axiom is presented in [4]. Another characterization using the notion of an M-graph is given in [3]. In this paper, we introduce a new path, namely a 'proper vertex sub path' for every u, v-path in a graph and discuss a characterization of the J-transit function satisfying the monotone axiom using this notion. In this approach, we obtain a modification of the concept of M-graph and a forbidden induced subgraph characterization of J-monotone graphs which improves all the other characterizations so far obtained.

In section 2, we survey the forbidden subgraph characterization of the J-monotone graphs using the Peano axiom. In section 3, we discuss the new characterization of the J-monotone graphs using the idea of the proper vertex subpath. In section 4, we characterize the graphs for which the transit function J satisfies the betweenness axiom $b2$.

We consider only connected, simple, loopless and finite graphs.

2. Monotone property of the induced path transit function

A transit function on a finite set V is a function $R : V \times V \to 2^V$ satisfying the three transit axioms

$$(t1) \quad u \in R(u, v), \text{ for any } u, v \in V,$$

$$(t2) \quad R(u, v) = R(v, u) \text{ for all } u, v \in V,$$

$$(t3) \quad R(u, u) = \{u\} \text{ for all } u \in V.$$

If G is a graph with vertex set V and R a transit function on V, then we say that R is a transit function on G.

The induced path transit function J of G, also known as the minimal path function or monophonic function, is defined as

$$J(u, v) = \{z \in V | z \text{ lies on some induced } u, v\text{-path in } G\}.$$

A set $W \subseteq V$ is an *R-convex* set if $R(u, v) \subseteq W$, for all u, v in W. The family C of R-convex sets in G is an abstract convexity, in the sense that, it is closed under intersections and nested union and both $\emptyset$ and V are convex sets. The family of R-convex sets is called the *R-convexity* on V. If G is a graph with vertex set V and if C is a convexity on V, then we say that C is a convexity on G.

The smallest R-convex subset containing a subset W of V is denoted by $\langle W \rangle_R$ or $\langle W \rangle$, if no confusion arises for R and is called the *R-convex hull* of W.

If $W = \{u, v\}$, we denote $\langle W \rangle$ by $\langle u, v \rangle$. If R is a transit function on V, then the transit function $R^* : V \times V \to 2^V$ defined by $R^*(u, v) = \langle u, v \rangle$, for $u, v \in V$, is called the segment transit function associated with R.

An abstract convexity C on a nonempty set V is said to be Join Hull Commutative (JHC) if for every $A \subseteq V$ and $x \in V$, $\langle x \cup A \rangle_C = \cup\{\langle x, a \rangle_C | a \in A\}$.

A transit function R is said to satisfy the *monotone axiom* if $x, y \in R(u, v)$ implies $R(x, y) \subseteq R(u, v)$. Or in other words, R is monotone if every $R(u, v)$ is R-convex.

A transit function R is said to satisfy the *Peano* axiom if, for any $u, v, w \in V, x \in R(v, w), y \in R(u, x)$, there is an $z\ inR(u, v)$ such that $y \in R(w, z)$.

We have a simple characterization of the J-monotone graphs using the idea of iteration number. If R is a transit function on $V(G)$, iteration number of an S subset of V with respect to R is defined as follows:

$$R(S) = \cup\{R(u, v)|u, v \in S\}.$$

$$R^2(S) = \cup\{R(u, v)|u, v \in R(S)\}. \text{ etc.,}$$

$$R^t(S) = \cup\{R(u, v)|u, v \in R^{(t-1)}(S)\}$$

Now the iteration number of S associated with the transit function R, denoted as $in(S)$ is defined as the smallest number k such that $R^k(S) = R^{(k+1)}(S) = \langle S \rangle$. The iteration number of G associated with R is defined as the supremum of $in(S)$ over all subsets S of $V(G)$. The iteration number of the J-transit function is called the minimal path iteration number in [3]. The following result is proved in [3].

> The induced path transit function J on a connected graph G is monotone if and only if the minimal path iteration number of G is one.

The following result shows that all connected graphs G with at most four vertices are J-monotone, see [4].

Theorem 1. *The induced path transit function J on any connected graph G containing four or less vertices satisfies the Peano axiom.*

Now, we discuss the results using the forbidden induced subgraphs, which are described in [4]. If G and H are two connected graphs, we say that G is H-free if G has no subgraph isomorphic to H as induced subgraph.

Proposition 1. *The induced path convexity (J-convexity) on a graph is JHC ([7]).*

Proposition 2. *If a transit function R satisfies the Peano axiom, then it is R-monotone ([25]).*

Proposition 3. *If R is a transit function, then the R-convexity is JHC if and only if R^* satisfies the Peano axiom ([25]).*

Proposition 4. *If R is a transit function with a JHC R-convexity, then R is R-monotone if and only if R satisfies the Peano axiom ([4]).*

The proof of the last Proposition follows directly from Propositions 2 and 3. Since the induced path convexity is JHC, we have the following corollary.

Corollary 1. *The induced path transit function J on a connected graph G is J-monotone if and only if it satisfies the Peano axiom ([4]).*

The two special graphs that is going to play a main role in this section are the following: **Subdivided $K_{2,3}$** and **subdivided $K_{2,3}$ with a chord**.

Let G_1 be the graph obtained by the subdivision of the edges of a $K_{2,3}$. Let u and v be the degree three vertices of G_1. Then there are three u, v-paths. Label them as P_s, P_t and P. Let s and t be the neighbors of u on P_s and P_t respectively. Let the neighbor of v on P_s be a'. Now allow a' to have adjacency with at least one interior vertex of P. The resulting graph is called a subdivided $K_{2,3}$ with a chord. The path P is called the central axis and the interior vertices of P are called the central vertices. The cycle formed by $P_s \cup P_t$ is called the cycle of the G_i. Since $K_{2,3}$ is planar and the subdivision graph of a planar graph is again planar, it follows that G_1 and G_2 are planar graphs.

We can easily see that the induced path transit function J on the graph $G_i, i = 1, 2$, does not satisfy the Peano axiom, since $v \in J(s, t)$, $y \in J(u, v)$, $J(u, t) = \{u, t\}$, but there is no $z \in J(s, u)$ with $y \in J(z, t)$, where y is any central vertex.

Theorem 2. *Let G be connected graph with at least five vertices. Then the induced path transit function J of G satisfies the Peano axiom only if it has no subdivided $K_{2,3}$ or subdivided $K_{2,3}$ with a chord as induced subgraph.*

Using the subgraphs $G_i, i = 1, 2$, a forbidden induced subgraph characterization is available only for planar graphs, see [4].

Notation 1. Suppose G has an induced subgraph isomorphic to a G_i, so that there is no induced path in G connecting the 's' and 't' vertices of the G_i and containing its 'f' vertex, then we say G has an induced $G_i(s, \hat{f}, t)$, for $i = 1, 2$.

Theorem 3. *The induced path transit function J on a connected planar graph G satisfies the peano axiom if and only if G is $G_i(s, \hat{f}, t)$ - free for $i = 1, 2$.*

3. A new characterization of J-monotone graph

We will first give the notations and terminologies that are used in this section. If a and b are two vertices of a path P, then the subpath of P from a to b is denoted by $a \to P \to b$. For a vertex u, u is a vertex on $a \to P \to b$ is denoted by $u \in a \to P \to b$ and u is not a vertex on $a \to P \to b$ is denoted by $u \notin a \to P \to b$. If $u \in a \to P \to b$, then the facts $u \neq a, u \neq b, u \neq a, b$ are denoted by $u \in (a) \to P \to b, u \in a \to P \to (b)$ and $u \in (a) \to P \to (b)$ respectively.

The new characterization of J-monotone graphs is based on the idea of a special type of subpath of a given path in a graph G known as **Proper vertex subpath of a path**, which is defined as follows:

Let G be a connected graph and P be a non trivial path in G. Then a path Q in G is said to be a proper vertex subpath of P if $V(Q) \subset V(P)$.
We can characterize the J-monotone graphs using this subpath.

Theorem 4. *The induced path transit function J on a connected graph with atleast 5 vertices is not J-monotone if and only if there exists vertices a, b, u, v and induced a, b paths P_1 and P_2 containing u and v respectively and an induced u, v path P of length ≥ 2 so that every a, b path containing an internal vertex of P has an a, b proper vertex subpath.*

Proof. Suppose the induced path transit function J on G is not J monotone. So there exists vertices a, b, u, v and z so that $u, v \in J(a,b)$, $z \in J(u,v)$ but $z \notin J(a,b)$. Hence there exist induced a, b paths P_1 and P_2 containing u and v respectively and an induced u, v path P containing z so that z is not on any a, b induced path. Let u' be the first vertex on $(z) \to P \to u$ lying on an a, b induced path P_1' and v' be the first vertex on $(z) \to P \to v$ lying on an a, b induced path P_2'. Relabel u', v', P_1' and P_2' as u_1, v_1, P_1 and P_2 respectively, we have a, b induced paths P_1 and P_2 containing u and v respectively so that no internal vertex in P lies on a, b induced path.

Suppose Q is an a, b path containing an internal vertex of P. Then Q is not an induced path. Hence there exists $a_1, b_1 \in V(Q)$ so that a_1 is adjacent to b_1 in G but $a_1b_1 \notin E(Q)$. So $Q_1 : a \to Q \to a_1 \to b_1 \to Q \to b$ is an a, b proper vertex subpath of Q.

Conversely, suppose there exist vertices a_1, b_1, u, v and induced paths P_1, P_2 and P on G satisfying the required conditions of the theorem. Clearly $u, v \in J(a,b)$ and if z is an internal vertex of P, then $z \in J(u,v)$. We can prove that $z \notin J(a,b)$. Let Q be any a, b path containing z, then Q has an a, b proper vertex subpath Q_1. Therefore $V(Q_1) \subsetneq V(Q)$. So there exist a vertex on Q which is not on Q_1. Suppose we are traversing from a to b along Q, let c_1 be the first vertex common to Q and Q_1 so that its immediate successor on Q is not on Q_1. Let c_2 be the first vertex after c_1 which is common to Q and Q_1. Then c_1 and c_2 are consecutive vertices of Q_1, hence c_1 is adjacent to c_2; but c_1c_2 is not an edge of Q. Hence Q is not an induced path. Thus there is no a, b induced path in G containing z. Therefore $z \notin J(a,b)$. This completes the proof. □

A characterization of the J-monotone graphs using the notion of *M-graph* is given in [3]. We give the definition of the M-graph and describe the characterization as presented in [3].

An *M-graph* is constructed as follows: Take two vertices a and b and connect them by two distinct paths (say) P_1 and P_2 of length at least two which can have common edges and vertices other than a and b. Choose an internal vertex u on P_1 which is not on P_2 and internal vertex v on P_2 which is not on P_1 and connect them by a path P of length greater than one and internally vertex disjoint with P_1 and P_2. Now allow the internal vertices of P_1 which are not on P_2 to have adjacency with some internal vertices of P_2 which are not on P_1. The graph thus constructed is called an *M-graph*. The internal vertices of P are called the internal vertices of M.

A connected graph G is said to contain an *M-graph* perfectly if

(i) G contains the *M-graph* as a subgraph.
(ii) The paths P_1, P_2 and P of the M- graph remain as induced paths in G.
(iii) Every $a - b$ path Q containing internal vertex z of P possesses a chord from $a \to Q \to (z)$ to $(z) \to Q \to b$.

We have the following characterization ([3]):

Theorem 5 ([3]). *The induced path transit function J on a connected graph G is monotone if and only if it does not contain any M-graph perfectly.*

Our aim is to obtain a forbidden induced subgraph characterization of J-monotone graphs. We now show that the perfect containment of an M-graph can be replaced by an induced subgraph which we call M_1. This can be described as follows.

We shall reserve the notations a, b, u, v for the specified vertices of M and the paths P_1, P_2 and P for the specified paths of M. We shall use terminologies like the a-vertex of M, the P-path of M and so on. Let G be a graph with at least five vertices. Suppose G has an induced subgraph isomorphic to an M-graph. Then the subgraph of G induced by all the vertices of the M-graph together with the vertices of all $a - b$ paths in G containing at least one internal vertex of P and possessing at least one proper vertex subpath is called an M_1-graph. Using this terminology Theorem 4 can be restated as follows.

Theorem 6. *The induced path transit function J on a connected graph G with at least three vertices is J monotone if and only if it is M_1-free.*

Proof. Suppose the induced path transit function J on G is not J- monotone. Then by Theorem 4, there exists vertices a, b, u, v and induced a, b paths P_1 and P_2 containing u and v respectively and an induced u, v path P of length at least two so that every a, b path containing an internal vertex of P has an a, b proper vertex subpath and the only vertices common to P and $P_1 \cup P_2$ are u and v. Then the induced subgraph of G induced by all the vertices of P_1, P_2, P together with the vertices of all a, b paths in G containing at least one internal vertex of P and possessing at least one proper vertex subpath is isomorphic to M_1, which is false.

Now suppose G contains an induced subgraph isomorphic to M_1. Let Q be a path connecting the a and b vertices of M_1 and containing an internal vertex z of the P path of M_1. Then Q has a proper vertex subpath Q_1 containing z. Since Q_1 is an $a - b$ path containing z, it has a proper vertex subpath Q_2 containing z. Proceeding like this we get a sequence of a, b subpaths, $Q_1, Q_2, \ldots,$ of Q such that each Q_i contains z and $V(Q_i) \subsetneq V(Q_{i-1})$; $i = 1, 2, \ldots$; $Q_0 = Q$. Hence after a certain stage we will arrive at an element Q_k of the sequence $Q_1, Q_2, \ldots,$ so that $a \to Q_k \to (z)$ and $z \to Q_k \to b$ are induced paths. Hence Q_k has no proper vertex subpath containing z. This conflict shows that there is no induced a, b path containing z. This shows that z is not an element of $J(a, b)$. Hence G is not J-monotone. □

Remark 1. Suppose we have defined the proper vertex subpaths. Then we can define the perfect containment of an M-graph as follows. A connected graph G is said to contain an M-graph perfectly if

(i) G contains the M-graph as an induced sub graph.
(ii) Every a, b path Q containing an internal vertex z of P has a proper vertex subpath.

Remark 2. Using Remark 2, we can see that a perfectly contained M-graph and an induced M_1 subgraph of a connected graph G are the same. Using this terminology Theorem 6 is nothing but Theorem 5.

Remark 3. The subdivided $K_{2,3}$ and the subdivided $K_{2,3}$ with a chord are M-graphs.

4. Betweenness and monotone axioms

Transit functions are introduced in [16] to study three basic notions in metric graph theory, namely the interval, convexity and betweenness. The transit function R highlights the elements w that might be "between" u and v. So it may play a role in the study of betweenness. A transit function R is a betweenness if it satisfies the two betweenness axioms:

$$(b1) \quad x \in R(u, v) \Rightarrow u \notin R(x, v), \text{ for } x \neq u.$$

$$(b2) \quad x \in R(u, v) \Rightarrow R(u, x) \subseteq R(u, v).$$

The induced path transit function is not a betweenness on every graph. It is shown in [18] that the induced path transit function J satisfies the betweenness axiom $(b1)$ and hence is a betweenness if and only if the graph is HHD-free. The identification of graphs for which the J transit function satisfies the betweenness axiom $(b2)$ has not been solved yet. We can easily see that if a transit function R satisfies the monotone axiom, then R satisfies $(b2)$, but not conversely. In this section, we identify the graphs for which J satisfies the $(b2)$ axiom, but not monotone. The smallest graph for which J satisfies $(b2)$, but not monotone is the complete bipartite graph $K_{2,3}$. The family of graphs for which this holds can be constructed from $K_{2,3}$. The construction is as follows.

Let u and v be the degree three vertices of a $K_{2,3}$. Let p, q and z be its degree two vertices. Subdivide the path $u \rightarrow z \rightarrow v$ and make p adjacent with some internal vertices of $u \rightarrow z \rightarrow v$ so that the neighbors of u and v on $u \rightarrow z \rightarrow v$ on the subdivided graph are adjacent to p. Let us denote the resulting graph by $K_{2,3}*$.

An immediate observation is that the induced path transit function J on $K_{2,3}*$ and $K_{2,3}$ satisfies the $b2$ axiom but not the monotone axiom. The following are some of the graphs in this class. There are two graphs on five vertices and 2^{n-6} graphs on $n \geq 6$.

The following proposition shows that these are the only graphs for which the induced path transit function satisfies $b2$ but not the monotone axiom.

Proposition 5. *Let G be a connected graph which is $K_{2,3}$ and $K_{2,3}*$-free. Then the induced path transit function J on G satisfies the $b2$ axiom if and only if it is monotone.*

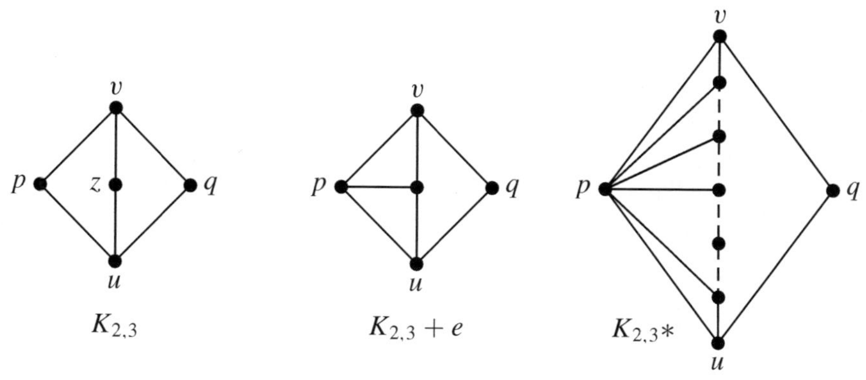

Figure 1.

Proof. The necessary part is clear. Let us prove the sufficiency part. Suppose the induced path transit function J on G satisfies $x \in J(u, v) \implies J(u, x) \subseteq J(u, v)$ for all $u, v, x \in V(G) \dots (m)$. We can prove that $J(u, v) \subseteq J(a, b)$ for all $u, v \in J(a, b) \in V(G)$, for any two vertices a and b of G. Assume the contrary. So there exists $a, b, u, v, z \in V(G)$ with $u, v \in J(a, b), z \in J(u, v)$ but $z \notin J(a, b)$. Hence there exists $a-b$ induced paths P_1 and P_2 containing u and v respectively and a $u-v$ induced path Q containing z. Since $z \notin J(a, b)$ the length of Q is at least two and P_1 and P_2 are different. Also we can choose the paths P_1 and P_2 and the vertices u and v so that no internal vertex on $u \to Q \to v$ belongs to $J(a, b)$. Let z_1 and z_2 be the neighbors of u and v respectively on Q. By (m), we have $z_1, z_2 \notin J(a, u) \cup J(b, u) \cup J(a, v) \cup J(b, v)$. Now $a \to P_1 \to u \to Q \to v$ is an $a - v$ path containing z_2. Hence there exists a chord from an internal vertex of $a \to P_1 \to u$ to v. Let p be the internal vertex on $a \to P_1 \to u$ nearest to u. Now $a \to P_1 \to p \to v \to Q \to u$ is an $a - u$ path containing z_1. Hence p is adjacent to u. By a similar argument we can establish the existence of an internal vertex q on $b \to P1 \to u$ which is adjacent to both u and v. In this case we can see that if an internal vertex of $u \to Q \to v$ is adjacent to p then no internal vertex of $u \to Q \to v$ is adjacent to q and vice versa; since no internal vertex on $u \to Q \to v$ belongs to $J(a, b)$. Hence one of the two cycles $C_1 : u \to Q \to v \to p \to u$; $C_2 : u \to Q \to v \to q \to u$ is an induced cycle. Both cycles are of the same length with length at least four. Assume that C_2 is an induced cycle. Since the graphs in figure 1 are forbidden, the length of C_1 and C_2 cannot be four. Our next claim is that both z_1 and z_2 are adjacent to p. If none are adjacent to p, we get the contradiction $v \in J(z, q), p \in J(z, v)$ but $p \notin J(z, q)$. If one of them say z_2 is adjacent to p and z_1 is not adjacent to p, we get the same contradiction. This shows that the subgraph induced by the vertices p, q, u, v and the path Q is either isomorphic to $K_{2,3}$ or $K_{2,3}*$. Since both $K_{2,3}$ and $K_{2,3}*$ are forbidden, the proposition follows. □

Acknowledgement

The authors thank Professor Pierre Duchet for his valuable ideas and discussions during the preparation of the last section on betweenness axiom $(b2)$.

References

[1] J. Calder, Some elementary properties of interval convexities, *J. London Math. Soc.*, **3** (1971) 422–428.

[2] M. Changat, S. Klavžar and H. M. Mulder, The All paths transit function of a graph, *Czech. Math. J.*, **51** (2001) no. 126, 439–448.

[3] M. Changat and J. Mathew, Interval monotone graphs: minimal path convexity. Proc. of the conf. on graph connections (eds.) R. Balakrishnan, H. M. Mulder and A. Vijayakumar, New Delhi (1999) 87–90.

[4] M. Changat and J. Mathew, Induced path transit function, monotone and Peano axioms, *Discrete Math.*, **286** (2004) 185–194.

[5] V. D. Chepoi, d-convex sets in graphs, Ph.D. dissertation, Moldova State University, Kishinev (1986) (Russian).

[6] F. F. Dragon, F. Nicolai and A. Brandstadt, Convexity and HHD-free graphs, *SIAM. J. Discrete Math.*, **12** (1999) no. 1, 119–135.

[7] P. Duchet, Convex sets in graphs II. Interval convexities (unpublished) (1984).

[8] P. Duchet, Convexity in combinatorial structures, *Rend. Circ. Mat. Palermo (2) Suppl.*, **14** (1987) 261–293.

[9] P. Duchet, Convex sets in graphs II. Minimal path convexity, *J. Combin. Theory Ser. B*, **44** (1988) 307–316.

[10] P. Duchet, Discrete convexity: retractions, morphisms and partition problem. Proc. of the conf. on graph connections (eds.) R. Balakrishnan, H. M. Mulder and A. Vijayakumar, New Delhi (1999) 10–18.

[11] M. Farber and R. E. Jamison, Convexity in graphs and hypergraphs, *SIAM J. Algebraic Discrete Methods*, **7** (1986) 433–444.

[12] M. Farber and R. E. Jamison, On local convexity in graphs, *Discrete Math.*, **66** (1987) 231–247.

[13] R. E. Jamison and Waldner, A perspective on abstract convexity: classifying alignments by varieties, in convexity and related combinatorial geometry. Proceedings of 2nd Oklahoma conf. (eds.) D. C. Kay and M. Breen, New York (1982) 113–150.

[14] S. Klavžar and H. M. Mulder, Median graphs: characterizations, location theory and related structures, *J. Combin. Math. Combin. Comp.*, **30** (1999) 103–127.

[15] H. M. Mulder, The Interval Function of a Graph. Mathematical Centre Tracts 132, Mathematisch Centrum, Amsterdam (1980).

[16] H. M. Mulder, Transit functions on graphs (and posets), (in this volume) 117–130.

[17] M. Mollard, Interval regularity does not lead to interval monotonocity, *Discrete. Math.*, **118** (1993) 233–237.

[18] M. A. Morgana and H. M. Mulder, The induced path convexity, betweenness and svelte graphs, *Discrete Math.*, **254** (2002) 349–370.

[19] L. Nebeský, Characterization of the set of all shortest paths in a connected graph, *Mathematica Bohemica*, **119** (1994) 15–20.

[20] L. Nebeský, Characterization the interval function of a connected graph Czec, *Math. Journal*, **44** (1994) 173–178.

[21] K. Nomora, A remark of Mulder's conjucture about interval regular graphs, *Discrete. Math.*, **147** (1995) 307–311.

[22] E. Sampathkumar, Convex sets in graphs, *Indian J. Pure Appl. Math.*, **15** (1984) 1065–1071.

[23] V. P. Soltan, d-convexity in graphs, *Soviet Math. Dokl.*, **28** (1983) 419–421.

[24] V. P. Soltan and V. D. Chepoi, Conditions for invariance of set diameters under d-convexification in a graph, *Cybernetics*, **19** (1983) 750–756.

[25] M. L. J. van de Vel, Theory of Convex Structures, North Holland, Amsterdam (1993).

Joint Proc. Int. Wrk.shps. Conv. in Discr. Struct. and Metric & Conv. Graph Th. No. 5, 2008, pp. 57–71.

Radon and Helly Numbers of Segment Spaces

Pierre Duchet

CNRS, Paris
Combinatoire & Optimisation, Univ. Paris 6, Case 189, 4 place Jussieu, 75252 Paris Cedex 05, France
e-mail: Pierre.Duchet@math.jussieu.fr

Abstract. In the special case of *segment spaces* (convexity spaces whose Carathéodory number c equals 1 or 2) we answer a problem raised in 1971 by Kay and Womble: we prove that positive integers h and r are respective Helly and Radon numbers of some convexity space with $c = 1$ (resp. $c = 2$) iff $r = h + 1$ (resp. $h < r \leq \lfloor 3h/2 \rfloor + 1$). For $c = 2$ and $h \geq 2$ all admissible values of h and r can be realized by a finite convex geometry which is the geodesic convexity of some metric space.

Keywords. Convexity space, segment convexity, interval convexity, Helly number, Radon number, Carathéodory number, transversal, node-covering number, hypergraph, geodesic, metric space, graph convexity, convex geometry, antimatroid.

2000 Subject Classification: 05D15, 05C70, 05C65, 51D99, 06A15, 52A01

1. Introduction

Classical theorems due to Carathéodory (1907), Radon (1921) and Helly (1923) express typical combinatorial properties of ordinary convex sets in $\mathbb{R}^d$. They have led to numerous refinements, developments and applications (see [1], [11]). As soon as some convex hull operator $\langle\cdot\rangle$ is available, the statement of those theorems make sense in the pure set-theoretic setting of *abstract convexity*.

According to a suggestion by Motzkin [21] (see also Levi [19]), axiomatic convexity is nowadays developed as a theory of algebraic closure systems (see [5,6,16]). A *convexity space* is a pair $(X, \mathscr{C})$ formed with a ground-set X (whose elements are named *points*) equipped with a collection $\mathscr{C}$ of subsets (the *convex sets*) that is preserved under arbitrary intersection and arbitrary upward-nested union. Note that both $\emptyset$ and X are convex sets. The *convex hull* of an X-subset A, denoted by $\langle A\rangle_{\mathscr{C}}$ or by $\mathscr{C}(A)$, is nothing else but the intersection of all convex sets containing A. The axiom on nested unions appears to be equivalent ([5], cf. [16,24]) to the following property, known as *domain finiteness*:

(DF) For any $A \subseteq X$, $\langle A\rangle_{\mathscr{C}} = \cup_{F\subseteq A, F\text{finite}}\langle F\rangle_{\mathscr{C}}$.

To any convexity space can be attached three combinatorial parameters. The Carathéodory number $c(\mathscr{C})$ (respectively the Helly number $h(\mathscr{C})$, the Radon number $r(\mathscr{C})$) is defined as the smallest integer k (if any) that fulfills the following property (C_k) (resp. (H_k), (R_k)):

(C$_k$) For any $A \subseteq X$, we have $\langle A\rangle_{\mathscr{C}} = \cup_{F\subseteq A, |F|\leq k}\langle F\rangle_{\mathscr{C}}$
(H$_k$) If any k members of a finite family $\mathscr{H}$ of convex sets share a common point, then all members of $\mathscr{H}$ share a common point.
(R$_k$) Every point set $A \subseteq X$ with at least k points admits a partition $A = A_1 \sqcup A_2$ such that $\langle A_1\rangle_{\mathscr{C}} \cap \langle A_2\rangle_{\mathscr{C}} \neq \emptyset$.

Following the pioneering idea of Levi's inequality $h(\mathscr{C}) + 1 \leq r(\mathscr{C})$ holds in every convexity space [19]), Kay and Womble [18] were the first to systematically investigate the relationships between those combinatorial parameters. They showed that, in every convexity space with a finite Carathéodory number c and a finite Helly number h, the Radon number r is finite and satisfies $r \leq ch + 1$. This inequality was later improved to $r \leq c(h-1) + 2$ by Eckhoff and Jamison (see [23] or [24]). Kay and Womble asked whether such inequalities, together with Levi's, would suffice to characterize all the possible values of parameters c, h and r for general convexity spaces (Problem 6 in [6]).

The present paper answers a special case of the question, namely that of convexity spaces with Carathéodory number at most 2. In the sequel, such spaces will be referred to as *segment convexity spaces* (in short, *segment spaces* or *2-spaces*): in a segment convexity, the convex hull of any set A is the union of segments with end points in A.

Two examples of segment spaces appearing in the literature deserve mention. In [17] Jamison-Waldner prove the validity of the famous partition conjecture (not discussed here, see [12]) for convexity spaces whose copoints have a special intersection property, named CIP(3,2) (a *copoint* attached to a point x is any inclusion-maximal convex set that avoids x; a space is CIP(3,2) if among any 3 distinct copoints attached to the same point, 2 of them are disjoint). CIP(3,2)-spaces include trees and order-convexities (where a subset is convex when it contains every element situated *between* – a self-explaining term– two of its elements). A result of Van de Vel [24] shows that convexity spaces with property CIP(3,2) are segment spaces. The Radon number of such spaces is in general one unit larger than the Helly number, the only exception consisting in CIP(3,2)-spaces with Helly number 2 and Radon number 4 (cf. [17]).

A second example of a segment convexity is supplied by the so-called *monophonic convexity* associated to a (connected) graph G: a set C is monophonically convex when it contains the vertices of every chordless path joining two vertices of C. We proved in [7] that the Carathéodory number of the monophonic convexity is 1 if G is complete, and 2 otherwise. As in the previous example, the Radon number is low: it exceeds the Helly number by one unit except if G is a tree but not a path (in such a tree, the Helly number is 2 and the Radon number is 4).

In contrast with the above examples, the present paper shows that the Radon number of a segment convexity with Helly number h can actually take all integer values from $h+1$ up to $\lfloor 3h/2\rfloor + 1$. The major part of the paper is devoted to prove that a set S of $\lfloor 3h/2\rfloor + 1$ points in a segment space with Helly number h admits a *Radon point*, i.e. a point that belongs to the convex hulls of two disjoint S-subsets. To that purpose, we make a repeated use of the *core* operator $A \mapsto \heartsuit A = \cap_{a\in A}\langle A - a\rangle$ (note that, by the definition of the Helly number, any set with more than h points has a non-empty core): the leading idea is to use the core-operator to find a set $R \subseteq S \cup \heartsuit S$ which can play the role of a 'Radon-point supplier' for S, i.e. such that every point of $\heartsuit R$ is a Radon point for S. If such an R with more than h points can be found, then $\heartsuit R$ is not empty and we are done.

After some terminological preliminaries, given in section 2, section 3 provides Radon-point suppliers with the help of '*S-triads*', i.e. those 3-point S-subsets that admits a non empty core; a compatibility condition imposes us to reject from S certain undesirable points, at a rate of at least 2 per S-triad. In other terms, the undesirable points of S should constitute a '2-transversal' of S-triads that fulfills a certain 'efficiency condition'. Dealing with arbitrary triple systems, section 4 provides a not too large efficient 2-transversal (a result that might be of independent interest in hypergraph theory). Inequality $r \leq \lfloor 3h/2 \rfloor + 1$ for segment convexities is derived in section 5 which gather the main results, including the straightforward case of convexity spaces with Carathéodory number 1. An explicit construction shows that the inequality is sharp: for every integers h and r such that $0 < h < r \leq \lfloor 3h/2 \rfloor + 1$, a finite segment-convexity space with Helly number h and Radon number r is exhibited. When $h \geq 2$, this space reveals to be both a *geodesic convexity space* and *a convex geometry*. We conclude with some remarks and open questions.

2. Preliminaries

Symbols $\sqcup$ and $\setminus$ mark respectively disjoint union and set difference. The addition (resp. deletion) of a single element set $\{x\}$ is marked as $A + x$ (resp. $A - x$).

We make use, especially in section 4, of the basic standard terminology relative to graphs and hypergraphs, for which [14] constitute a reliable reference. By a *t-graph* we mean a t-uniform simple hypergraph, i.e. a pair $(S, \mathscr{T})$ formed with a ground set S, whose elements are the *vertices*, and a collection $\mathscr{T}$ of t-element S-subsets, called the *edges*. We also say for short that $\mathscr{T}$ is a t-graph *on* S. A 2-graph is nothing else but a (simple) graph; 3-graphs are also known as *triple systems*. Given a t-graph $\mathscr{T}$, a set W that shares at least k vertices with any edge of $\mathscr{T}$ is called a *k-transversal* of $\mathscr{T}$ (cf [8]). The maximum number of pairwise disjoint $\mathscr{T}$-edges is the *matching number* of $\mathscr{T}$, usually denoted as $\nu(\mathscr{T})$.

In a convexity space $(X, \mathscr{C})$, the *convex hull* of a point set A is denoted by $\langle A \rangle_{\mathscr{C}}$ or simply by $\langle A \rangle$ when no doubt exists about the ambient convexity; the notation $\langle \{a, b, c, \dots\} \rangle$ is abridged into $\langle a, b, c, \dots \rangle$. A relation such as $x \in \langle A \rangle$ is called a *convex dependence*; alternatively, we say that x is *convexly generated* by A. The dependence is said to be *trivial* when $x \in A$ and to be *strict* otherwise.

By virtue of domain finiteness (viz. axiom (DF) in the introduction), any convexity space $(X, \mathscr{C})$ can be *presented* by ('generated by' or 'defined by' are alternative terms) a list $(\Omega_i : X^{k_i} \to 2^X)_i$ of finitary operators. Convex sets are precisely those X-subsets that are closed under all Ω_i's. When all the defining operators have arity 1 or 2, $\mathscr{C}$ is called an *interval-convexity* (cf. Calder [4]). As easily seen, any interval convexity $(X, \mathscr{C})$ can be presented by the following single binary operator, called the *segment* operator, defined for all $a, b \in X$ by

$$\sigma(a, b) = \langle a, b \rangle_{\mathscr{C}} \tag{1}$$

We say that a and b form a pair of *endpoints* of the segment $\langle a, b\rangle_{\mathscr{C}}$. Note that equality $a = b$ is permitted and that, in general, many different pairs of endpoints can generate the same segment.

To any metric space (X, d) can be associated an interval convexity space as follows. A point x is said to lie in *between* points a, b provided:

$$\mathrm{d}(a, b) = \mathrm{d}(a, x) + \mathrm{d}(x, b) \tag{2}$$

If, additionally, $x \neq a, b$, then x is said to be *strictly between* a and b. The *geodesic interval function*, defined by $\mathrm{I_d}(a, b) = \{x \in X; x \text{ lies in between } a \text{ and } b\}$ generates what is called the *geodesic* (or *metric*)*convexity* of the metric space (X, d). An easy instance of a geodesic convexity is supplied by the usual Euclidean convexity in $\mathbb{R}^d$. Other typical instances are *graph geodesic convexities*, i.e. geodesic convexities associated to connected graphs when considered as metric spaces: a set C of vertices is geodesically convex if it contains all the vertices of all shortest path having their endpoints in C. As easy examples show, the interval function $\mathrm{I_d}$ of a metric space generally differs from the corresponding segment operator.

An *extreme* point of a convex set C is any point $x \in C$ such that $C - x$ is convex. When, in a convexity space, every convex set coincides with the convex hull of its extreme points (Krein-Milman property), the space is named a *convex geometry.* Also known as 'antimatroids' or 'dual shelling structures', such convexities can be characterized by a single 'anti-exchange axiom' (see [13]) and have led to numerous developments in mathematical social science.

Classical combinatorial invariants attached to a convexity space $(X, \mathscr{C})$ can be viewed as expressing a certain degree of independence for its point sets.

Convex dependence. A point set A is said to be convexly independent if $a \notin \langle A - a\rangle$ for every $a \in A$, and *convexly dependent* otherwise. The maximum size of a convexly independent set in $(X, \mathscr{C})$ is the *convex rank*, denoted as $\mathrm{rank}(\mathscr{C})$.

Helly independence. The set $\bigcap_{a\in A}\langle A - a\rangle$, usually called the *(Helly-) core* of A and is denoted as $\heartsuit_{\mathscr{C}}A$, or simply as $\heartsuit A$ if no ambiguity can occur. Elements of $\heartsuit A$ are named the *core-points for A*. A point set is *Helly-independent* if it has an empty core, and *Helly dependent* otherwise. The maximum cardinality of a Helly-independent set in $(X, \mathscr{C})$ is the *Helly number* of $\mathscr{C}$, denoted by $h(\mathscr{C})$. This terminology agrees with the alternative definition given in the introduction, since it can be shown (see [3,4,22]) that $h(\mathscr{C}) \leq k$ holds if and only if convex sets in $\mathscr{C}$ satisfy condition (H_k).

Radon independence. An unordered pair of disjoint sets A_1 and A_2 such that $\langle A_1\rangle \cap \langle A_2\rangle \neq \emptyset$ is named a *Radon partition* of the union set $A_1 \sqcup A_2$. More generally we say that $\{A_1, A_2\}$ is a Radon partition *for* any set A that contains $A_1 \sqcup A_2$ or, equivalently, that A *admits* $\{A_1, A_2\}$ as a Radon partition. Every point in $\langle A_1\rangle \cap \langle A_2\rangle$ is called a *Radon point for A*. The set of all Radon points for A (corresponding to all possible Radon partitions) will be denoted by ✠A (to read as 'cross A'). A set is *Radon dependent* (or *2-partitionable*) if it admits a Radon partition and is said to be (resp.) *Radon independent* otherwise. The *Radon dimension* of a convexity space $(X, \mathscr{C})$ is the maximum size of a Radon-independent point set. Traditionally, the term ‘Radon *number*’, with notation $r(\mathscr{C})$, refers to the Radon dimension augmented by one unit (except in [24], where the ‘Radon number’ coincides with our ‘Radon dimension’); for instance if the whole space X is Radon independent, the Radon number equals $\mathrm{card}(X) + 1$. Levi's inequality $h(\mathscr{C}) + 1 \leq r(\mathscr{C})$ expresses that the straightforward inclusion $\maltese A \subseteq \heartsuit A$, holds for every point set A.

Carathéodory independence. A point set A is said to be *irredundant*, or *Carathéodory independent*, if $\langle A \rangle \not\subseteq \cup_{a \in A} \langle A - a \rangle$, and *redundant*, or *Carathéodory dependent*, otherwise. The *Carathéodory number* $c(\mathscr{C})$ of a convexity space $(X, \mathscr{C})$ is the maximum size of an irredundant point set. It is not difficult to see that this definition coincides with the one given in the introduction : we have $c(\mathscr{C}) \leq k$ iff, for any point set A, any member of the convex hull $\langle A \rangle_{\mathscr{C}}$ is convexly generated by no more than k points in A.

Segment-spaces. We observe that $c(\mathscr{C}) \leq 2$ iff the convex hull of any $A \subseteq X$ is the union of segments generated by pairs of points in A: $\langle A \rangle_{\mathscr{C}} = \sigma(A \times A)$. In other terms, a convexity with Carathéodory number at most 2 is an interval-convexity where a single application of the segment operator suffices to generate a convex hull (according to [15], $\mathscr{C}$, presented by σ, has ‘generating iteration number’ 1). This explains the term ‘*segment-convexities*’ we use to designate such convexities.

Remark 2.1. The Radon, Helly and convex independence is a hereditary notion: any subset of an independent set is independent too. This does not stand any more for Carathéodory independence (even in interval convexities). Note that Radon independence implies Helly independence, which in turn implies convex independence. Carathéodory independence also implies convex independence but is not logically related to Helly or Radon independence.

3. Cores and triads

The goal of this section is to associate large Helly-independent sets to a given Radon-independent point set in a segment convexity space. We begin with some general properties concerning convex hulls and cores. The next two lemmas will frequently be of implicit use; their easy proof is left to the reader.

Lemma 3.1. *In any convexity space $(X, \mathscr{C})$, the convex hull operator $S \in 2^X \mapsto \langle S \rangle_{\mathscr{C}}$ and the core operator $S \in 2^X \mapsto \heartsuit_{\mathscr{C}} S$ are isotone operators: for every $A, B \subseteq X$, the inclusion $A \subseteq B$ implies $\langle A \rangle_{\mathscr{C}} \subseteq \langle B \rangle_{\mathscr{C}}$ and $\heartsuit_{\mathscr{C}} A \subseteq \heartsuit_{\mathscr{C}} B$.* □

Lemma 3.2. *For any point set S in a convexity space we have $S \cup \heartsuit S \subseteq \langle S \rangle$, with $S \cap \heartsuit S = \emptyset$ iff S is convexly independent.* □

Lemma 3.3 (Double Core Formula). *Any point set S in a convexity space satisfies the following formula:*

$$\heartsuit(S \cup \heartsuit S) = \heartsuit S \tag{3}$$

Proof. By the definition of a core, for every $s \in S$ we have $\heartsuit S \subseteq \langle S - s\rangle$; hence, since $S \subseteq \langle S\rangle$, we have $(S \cup \heartsuit S) - s \subseteq \langle S - s\rangle$. Therefore, applying Lemma 3.1:

$$\heartsuit(S \cup \heartsuit S) = \cap_{s \in S \cup \heartsuit S}\langle (S \cup \heartsuit S) - s\rangle \subseteq \cap_{s \in S}\langle S - s\rangle = \heartsuit S$$

The reverse inclusion is straightforward, by Lemma 3.1. □

In segment spaces, Helly-dependent triples play an important role.

Definition 3.4. Let S be a point set in an arbitrary convexity space. Any 3-element S-subset which is Helly dependent is called a *triad* in S or an *S-triad.* When T is such a triad, any of its core-point $\theta \in \heartsuit T$ is called a *triadic point for S* (or an *S-triadic point*), and we say that θ is *supported by T*. The *triadic core* of S, denoted by $\overset{3}{\heartsuit} S$, is the set of all S-triadic points.

Lemma 3.5. *For any point set S in a segment space, holds:* $\heartsuit S = \overset{3}{\heartsuit} S \cup \maltese S$.

Proof. Obviously, S-triadic points and Radon points for S are core-points for S. Thus, it suffices to establish the inclusion $\heartsuit S \setminus \maltese S \subseteq \overset{3}{\heartsuit} S$. Let ξ a point in $\heartsuit S \setminus \maltese S$. Since we are in a segment space, ξ belongs to some segment $\langle x, y\rangle$ where x and y are possibly equal points of S. we have $\xi \in \langle S - x\rangle$ hence $\xi \in \maltese S$, contrarily to our hypotheses. Thus $x \neq y$. Since ξ is a core-point for S, we have $\xi \in \langle S - x\rangle \cap \langle S - y\rangle$, hence there exist points y', z in $S - x$ and x', z' in $S - y$ such that $\xi \in \langle y', z\rangle \cap \langle x', z'\rangle$. Since ξ is not a Radon point for S, we have (i) $\{x, y\} \cap \{y', z\} \neq \emptyset$, (ii) $\{x, y\} \cap \{x', z'\} \neq \emptyset$ and (iii) $\{y', z\} \cap \{x', z'\} \neq \emptyset$. By (i) and (ii), we may assume, without loss of generality, $y' = y$ and $x' = x$. Then (iii) implies $z' = z$. So, x, y and z are all distinct and we have $\xi \in \langle x, y\rangle \cap \langle y, z\rangle \cap \langle z, x\rangle$; hence ξ is a triadic point for S, as required. □

The next lemma is a crucial tool to obtain Radon-dependencies with the use of triadic points.

Lemma 3.6. *Let S be a point set in a segment convexity space. Then every S-triadic point that is supported by two distinct S-triads is a Radon point for S.*

Proof. We proceed by contradiction. Let $\xi \in \heartsuit T \cap \heartsuit T'$, where $T = \{x, y, z\}$ and $T' = \{x', y', z'\}$ are distinct S-triads and assume ξ is not a Radon point for S. Then $\xi \in \langle x, y\rangle \cap \langle x', y'\rangle$ implies $\{x, y\} \cap \{x', y'\} \neq \emptyset$ and we may assume, without loss of generality, $x = x'$. Similarly, $\xi \in \langle y, z\rangle \cap \langle y', z'\rangle$ implies $\{y, z\} \cap \{y', z'\} \neq \emptyset$ and we may assume $y = y'$. Now we have $\xi \in \langle x, z\rangle \cap \langle y', z'\rangle$, which implies $\{x, z\} \cap \{y', z'\} \neq \emptyset$; since $x = x'$, $x' \neq y', z'$, $z \neq y$ and $y = y'$, this forces $z = z'$ hence $T = T'$, a contradiction. □

The next lemma is a useful technical observation on convex hulls involving triadic points.

Lemma 3.7. *Let S be a point set in a segment space. If θ and θ' are S-triadic points (may be not distinct), respectively supported by S-triads T' and T'' (may be not distinct), then* $\langle \theta, \theta'\rangle \subseteq \maltese S \cup \heartsuit T' \cup \heartsuit T''$.

Proof. Assuming $\theta \in \heartsuit T'$ and $\theta' \in \heartsuit T''$ satisfy the hypotheses of the lemma, let $\xi \in \langle \theta, \theta' \rangle \setminus \maltese S$. Two easy observations can be made:

(i) $\xi \in \langle (T' - t') \cup (T'' - t'') \rangle$ for every $t' \in T'$, $t'' \in T''$
(ii) $\xi \notin \langle t', t'' \rangle$ for every $t' \in T'$, $t'' \in T''$.

Actually, (i) immediately follows from the definition of the cores $\heartsuit T'$ and $\heartsuit T''$, and (ii) follows from (i), since ξ is not a Radon point for S. Now suppose $\xi \notin \heartsuit T'$. Choose x in T' such that $\xi \notin \langle T' - x \rangle$ and let y be any point in T''. By (i), we have $\xi \in \langle (T' - x) \cup (T'' - y) \rangle$. Since the Carathéodory number is at most 2, the point ξ belongs to the convex hull of some 2-element subset of $(T' - x) \cup (T'' - y)$, say Z. Since $\xi \notin \langle T' - x \rangle$, (ii) implies $Z \subseteq T'' - y$, hence $\xi \in \langle T'' - y \rangle$. It follows that $\xi \in \langle T'' - y \rangle$ for every $y \in T''$, i.e. $\xi \in \heartsuit T''$, and the lemma is proved (notice that our proof never used $\theta \neq \theta'$ or $T' \neq T''$ as assumptions). □

Let us say that a point set R is a *Radon supplier* for another point set S if any core point of R is a Radon point for S. The next result shows that, in a segment space, any system of representatives of S-triad cores constitutes a Radon point supplier for S.

Proposition 3.8. *Let S be a point set in a segment space and let $\mathscr{T}$ denote the set of all S-triads. For each $T \in \mathscr{T}$, let θ_T denote some triadic point supported by T and put $\Theta = \{\theta_T; T \in \mathscr{T}\}$. Then $\heartsuit\Theta \subseteq \maltese S$.*

Proof. We proceed by contradiction, assuming $\heartsuit\Theta \setminus \maltese S$ contains a point ξ. By the double core formula (3), ξ is a core point for S, hence by lemma 3.5, ξ is an S-triadic point. Let $T \in \mathscr{T}$ such that $\xi \in \heartsuit T$. By definition of a core, $\xi \in \heartsuit\Theta$ implies $\xi \in \langle \Theta - \theta_T \rangle$. Since the Carathéodory number is at most 2, we have $\xi \in \langle \theta_{T'}, \theta_{T''} \rangle$ for some triads $T', T'' \in \mathscr{T} - T$. Now, by lemma 3.7, at least one of the triads T' and T'' supports ξ. Therefore, since $\xi \in \heartsuit T$, ξ is supported by distinct triads, hence, by lemma 3.5, is a Radon point for S, contrarily to our assumption. □

We now proceed to improve Proposition 3.8 in order to exhibit larger Radon point suppliers.

Definition 3.9. *Given an arbitrary 3-graph $\mathscr{T}$ on a set S, an S-subset U is called* $\mathscr{T}$-compatible *if it fulfills the following conditions (where Δ denotes the symmetric difference):*

(i) $|U \cap T| \leq 1$ *for every* $T \in \mathscr{T}$
(ii) *for every* $T, T' \in \mathscr{T}$ *such that* $|T \cap T'| = 2$, *we have* $U \cap (T \Delta T') = \emptyset$

Proposition 3.10. *Let S be a point set in a segment space and let $\mathscr{T}$ denote the set of all S-triads. For each $T \in \mathscr{T}$, let θ_T denote some triadic point supported by T and put $\Theta = \{\theta_T; T \in \mathscr{T}\}$. Then, for any $\mathscr{T}$-compatible S-subset U, we have $\heartsuit(\Theta \cup U) \subseteq \maltese S$.*

Proof. As for proposition 3.8, we proceed by contradiction, assuming $\heartsuit(\Theta \cup U) \setminus \maltese S$ contains some point ξ. By the Double Core formula, we have $\xi \in \heartsuit T$ for some $T \in \mathscr{T}$. Then the hypothesis $\xi \in \heartsuit(\Theta \cup U)$ implies $\xi \in \langle (\Theta \cup U) - \theta_T \rangle$ hence $\xi \in \langle (\Theta - \theta_T) \cup U \rangle$. Therefore, ξ lies in some segment Z with endpoints in $(\Theta - \theta_T) \cup U$. Considering the set $Z \cap U$, three cases are to be examined; in each case a contradiction will occur.

Case (i). $Z = \{a, b\}$ where a and b are (possibly equal) points in U.

By condition (i) on $\mathscr{T}$-compatible sets (viz. definition 3.9), we have $|T \cap \{a, b\}| \leq 1$, hence $\xi \in \langle T\backslash\{a, b\}\rangle$ (for $\xi \in \heartsuit T$). Thus $\{\{a, b\}, T\backslash\{a, b\}\}$ constitutes a Radon partition for S with Radon point ξ. Contradiction.

Case (ii). $Z = \{a, \theta_{T'}\}$ where $a \in U$ and $T' \neq T$ (possibly $a = \theta_{T'}$).

Let us first suppose $|T \cap T'| \leq 1$. Then $\xi \in \langle T\backslash T'\rangle \cap \langle\{a\} \cup (T'\backslash T)\rangle$. It follows that $a \in T\backslash T'$ (otherwise S would admit $\{T\backslash T', \{a\} \cup (T'\backslash T)\}$ as a Radon partition with Radon point ξ). Hence $\xi \in \langle T - a\rangle$ (for $\xi \in \heartsuit T$). But now $\{T - a, \{a\} \cup (T'\backslash T)\}$ form a Radon partition for S, with Radon point ξ. Contradiction.

So, we may suppose $|T \cap T'| = 2$. Setting $T \cap T' = \{x, y\}$, we may assume, without loss of generality, $x \neq a$. We have $\xi \in \langle T - y\rangle$ (Since $\xi \in \heartsuit T$) and $\xi \in \langle\{a\} \cup T' - x\rangle$ (because $T' - x$ convexly generates $\theta_{T'}$ and $\{\theta_{T'}, a\}$ convexly generates ξ). Since ξ is not a Radon point for S,the sets $T - x$ and $\{a\} \cup T' - y$ meet. Since $T - x$ and $T' - y$ are disjoint sets, we obtain $a \in T - y$. Thus we have $a \neq x, y$ hence $a \in T \Delta T'$, a contradiction with condition (ii) on $\mathscr{T}$-compatible sets (viz. definition 3.9).

Case (iii). $Z = \{\theta_{T'}, \theta_{T''}\}$ where $T', T'' \neq T$ (possibly $T' = T''$).

Applying lemma 3.7, we obtain $\xi \in \heartsuit T' \cup \heartsuit T''$. Hence, the triadic point ξ, supported by T, is supported by two distinct S-triads. Then, Lemma 3.6 implies that ξ is a Radon point for S, a contradiction.

All possible cases having been treated, the proof is complete. □

4. Efficient transversals for triple systems

The goal of the present section is to establish the following theorem that will lead to a quantitative version of Proposition 3.10.

Theorem 4.1. *Let $\mathscr{T}$ be the 3-graph formed, in an arbitrary convexity space, with all triads of a point set S. Then S contains a $\mathscr{T}$-compatible subset with at least $\lceil 2s/3 \rceil - m$ points, where $s = |S|$ and $m = |\mathscr{T}|$*

The above statement is in fact of strict combinatorial nature: convexity plays no role in it.

Definition and Remark 4.2. Given a 3-graph $\mathscr{T}$ on a vertex set V, we say that a 2-transversal W of $\mathscr{T}$ is *efficient* if $V\backslash W$ is a $\mathscr{T}$-compatible set (see definition 3.9). Remark that, by definition 3.9, an efficient 2-transversal is nothing else but a 2-transversal that contains all singletons of the form $T\backslash T'$ where $T, T' \in \mathscr{T}$ and $|T \cap T'| = 2$; elements of such singletons are named *forced vertices* of $\mathscr{T}$.

Theorem 4.1 can be seen as a consequence of the following statement.

Theorem 4.3. *Every 3-graph with n vertices and m edges admits an efficient 2-transversal with no more than $\lfloor n/3 \rfloor + m$ elements.*

Proof of Theorem 4.1 via Theorem 4.3. Let $\mathscr{T}$ be the 3-graph formed in a convexity space with all triads of a point set S. Theorem 4.3 supplies an efficient 2-transversal

W for $\mathcal{T}$ with at most $\lfloor n/3 \rfloor + m$ points, where $n = |\cup \mathcal{T}|$ and $m = |\mathcal{T}|$; therefore $S \backslash W$ constitute a $\mathcal{T}$-compatible set with at least $s - \lfloor n/3 \rfloor - m$ points. Since $s \geq n$, we have $s - \lfloor n/3 \rfloor \geq \lceil 2s/3 \rceil$, hence $|S \backslash W| \geq \lceil 2s/3 \rceil - m$. □

To prove Theorem 4.3, we establish it in two particular cases, under more precise forms: such is the object of the next two lemmas.

Lemma 4.4. *Suppose $\mathcal{T}$ is a 3-graph with n vertices in which any edge shares 2 vertices with some other edge. Then $\mathcal{T}$ admits an efficient 2-transversal of size at most* $\min(3|\mathcal{T}|/2, \lfloor n/3 \rfloor + |\mathcal{T}|)$.

Proof. Denote by G_2 the 2-intersection graph of $\mathcal{T}$, i.e. the graph whose vertices are $\mathcal{T}$-edges, two $\mathcal{T}$-edges being adjacent in G_2 when they share exactly 2 vertices. Connected components of G_2 form a partition of $\mathcal{T}$ into what we name 2-components of $\mathcal{T}$. To obtain the existence of an efficient 2-transversal of size at most $3|\mathcal{T}|/2$, it suffices to establish that any 2-component $\mathcal{K}$ of $\mathcal{T}$ admits an efficient 2-transversal $W_{\mathcal{K}}$ with no more than $3|\mathcal{K}|/2$ elements. To that purpose, let us order the members of $\mathcal{K}$ as $T_1, T_2, \ldots, T_p$ (with $p \geq 2$) so that, for each $i = 1, 2, \ldots, p$, $\{T_1, T_2, \ldots T_i\}$ induces a connected subgraph of G_2. Now $W_{\mathcal{K}}$ is defined as follows: if every vertex of T_1 is a forced vertex, we set $W_{\mathcal{K}} = T_1 \cup T_2 \cup \cdots \cup T_p$; otherwise we set $W_{\mathcal{K}} = (T_1 \cup T_2 \cup \cdots \cup T_p) - x$, where x is an arbitrary unforced vertex of T_1.

Clearly, $W_{\mathcal{K}}$ is an efficient 2-transversal of $\mathcal{K}$. We claim that $|W_{\mathcal{K}}| \leq 3p/2$. Actually, we have $|T_1 \cup T_2 \cup \cdots \cup T_p| \leq |T_1| + p - 1 = p + 2$ since, by construction, each set of the form $T_i \backslash (T_1 \cup T_2 \cup \cdots \cup T_{i-1})$ has at most one element. If $p \geq 4$, we have $p + 2 \leq 3p/2$ and our claim follows. If $p = 2$ or 3, a (quick) exhaustive examination shows that, in all possible cases, T_1 contains some unforced vertex; hence $|W| \leq p+1$, henceforth $|W_{\mathcal{K}}| \leq 3p/2$, as required. Thus the inequality $|W_{\mathcal{K}}| \leq 3p/2$ holds for every $p \geq 2$.

Consequently, $W = \cup\{W_{\mathcal{K}} | \mathcal{K} \text{ is a } 2-\text{component of } \mathcal{T}\}$ constitutes an efficient 2-transversal of $\mathcal{T}$ with no more than $3|\mathcal{T}|/2$ elements. To conclude the proof of Lemma 4.4, it remains to observe that $|W| \leq 3|\mathcal{T}|/2$ is equivalent to $2|W|/3 \leq |\mathcal{T}|$ hence implies $|W| \leq |W|/3 + |\mathcal{T}| \leq n/3 + |\mathcal{T}|$. □

Let us recall that a hypergraph is *linear* when any two of its edges have at most one vertex in common. Since, obviously, a linear 3-graph has no forced vertex, each of its 2-transversal is efficient.

Lemma 4.5. *Every linear 3-graph $\mathcal{T}$ with n vertices, m edges and matching number ν admits a 2-transversal of size at most $\nu + m$, hence at most $\lfloor n/3 \rfloor + m$.*

Proof. By induction on m, the lemma being trivial for $m \leq 1$. Let $\mathcal{T}$ be a linear 3-graph with $m \geq 2$ edges and matching number ν. Suppose $\mathcal{T}$ contains some cycle of length $p \geq 2$, i.e. some sequence $x_1, T_1, x_2, T_2, \ldots, x_p, T_p, x_1$ where $x_1, x_2, \ldots, x_p$ are distinct vertices and $T_1, T_2, \ldots, T_p$ are distinct edges such that $x_i \in T_i \cap T_{i+1}$ for $i = 1, 2, \ldots, p$ (mod. p). The 3-graph $\mathcal{T}' = \mathcal{T} \backslash \{T_1, T_2, \ldots, T_p\}$ admits, by the induction hypothesis, a 2-transversal W' of size at most $\nu(\mathcal{T}') + |\mathcal{T}'|$; thus $W' \cup \{x_1, x_2, \ldots, x_p\}$ is a 2-transversal of $\mathcal{T}$ with size not exceeding $\nu(\mathcal{T}') + |\mathcal{T}'| + p \leq \nu(\mathcal{T}) + |\mathcal{T}|$, as expected. So we may assume now that $\mathcal{T}$ contains no cycle. *A fortiori*, $\mathcal{T}$ is a *balanced hypergraph* (since such hypergraphs are defined by the absence of cycles of odd length in any of its subhypergraph). The Berge-Las Vergnas theorem (see [2], or [20] for a short proof) asserts that $\mathscr{T}$ admits a 1-transversal W_1 with ν elements. Adding to W_1 a vertex $v \in T \backslash W_1$ for every $\mathscr{T}$-edge T such that $|T \cap W_1| \leq 1$, we obtain a 2-transversal for $\mathscr{T}$ with size not more than $\nu + |\mathscr{T}|$, as required. Since, obviously, $\nu \leq n/3$, the lemma follows. □

We now derive Theorem 4.3 from the preceding lemmas.

Proof of Theorem 4.3. Let $\mathscr{T}$ a 3-graph with n vertices and m edges. In order to construct an efficient 2-transversal, we split $\mathscr{T}$ into three parts.

First of all, we consider the set $\mathscr{A} = \{T \in \mathscr{T};\ (\exists T' \in \mathscr{T})|T \cap T'| = 2\}$. Applying lemma 4.4 to the 3-graph $\mathscr{A}$ and visualizing the end of its proof, we see that $\mathscr{A}$ admits an efficient 2-transversal W such that $|W| \leq |W|/3 + |\mathscr{A}|$.

Then, we consider the set $\mathscr{A}'$ of all $\mathscr{T}$-edges meeting W in exactly one vertex. Adding to W a vertex $v \in T \backslash W$ for every $T \in \mathscr{A}'$, we obtain a set W' which is an efficient 2-transversal of $\mathscr{A} \cup \mathscr{A}'$, with size at most $|W|/3 + |\mathscr{A}| + |\mathscr{A}'|$.

Lastly, we consider the collection $\mathscr{A}''$ of all the remaining $\mathscr{T}$-edges. By virtue of the definition of $\mathscr{A}$, we know that $\mathscr{A}''$ is a linear 3-graph that admits, by lemma 4.5, a 2-transversal W'' with $|W''| \leq \nu(\mathscr{A}'') + |\mathscr{A}''| \leq |\cup \mathscr{A}''|/3 + |\mathscr{A}''|$.

We observe that, by the definition of $\mathscr{A}$, all forced vertices of $\mathscr{T}$ are in W, hence in W'. Thus, $W' \cup W''$ constitutes an efficient 2-transversal of $\mathscr{T}$. By the definition of $\mathscr{A}'$, no edge in $\mathscr{A}''$ meet W, hence we have $|W| + |\cup \mathscr{A}''| \leq n$. Therefore, summing our inequalities relative to W' and W'', we obtain $|W' \cup W''| \leq n/3 + |\mathscr{T}|$. Theorem 4.3 follows. □

5. Radon number vs Helly number

We are now ready to state and prove our main results.

Theorem 5.1. *Positive integers h and r are respective Helly and Radon numbers of some convexity space with Carathéodory number 1 if and only if $r = h + 1$.*

Proof. Recall the three general inequalities we presented in our introduction, due respectively to Levi, to Kay and Womble, and to Eckhoff and Jamison. By Levi's inequality, we have $r \geq h + 1$ while, when the Carathéodory number is 1, the two other inequalities reduce to $r \leq h + 1$. The 'only if' part of the theorem follows.

To prove the converse assertion, we construct a convexity space $(X_1(h), \mathscr{C}_1(h))$ with Carathéodory number 1 and Helly number h for every integer h. We set $X_1(h) = \cup_{i=1,\dots,h}\{t_i, \theta_i\}$ and define $(X_1(h), \mathscr{C}_1(h))$ as the convexity space presented by a single unary operator J_1, defined, for $i = 1, \dots, h$ by:

$$\mathrm{J}(t_i) = \{\theta_i\} \mathrm{and} \mathrm{J}_1(\theta_i) = \emptyset \tag{4}$$

In this convexity space, the only non trivial convex dependencies have the form $\theta_i \in \langle t_i \rangle$ for $i = 1, \dots, h$. It readily follows that the Carathéodory number is 1 and that the set $S = \{t_1, \dots, t_h\}$ is Helly-independent. To see that the Helly number is precisely h, observe that every $h+1$-point subset of $X_1(h)$ contains some $\{t_i, \theta_i\}$, hence is convexly dependent, and *a fortiori* Helly dependent. □

Theorem 5.2. *Positive integers h and r are respective Helly and Radon numbers of some convexity space with Carathéodory number 2 if and only if $h < r \le \lfloor 3h/2 \rfloor + 1$.*

Proof. As above, we begin with the 'only if part'. Suppose S is a Radon independent set. Denoting by $\mathcal{T}$ the set of all S-triads, we put $|\mathcal{T}| = m$. Choosing, for each $T \in \mathcal{T}$, a point $\theta_T \in \heartsuit T$, we obtain a set $\Theta = \{\theta_T\}_{T \in \mathcal{T}}$. By Theorem 4.1 there exists a $\mathcal{T}$-compatible S-subset U with at least $\lceil 2|S|/3 \rceil - m$ points. By Proposition 3.10, $\Theta \cup U$ has an empty core, i.e. is Helly-independent. Hence $|\Theta \cup U| = \lceil 2|S|/3 \rceil \le h$, or, equivalently, $|S| \le \lfloor 3h/2 \rfloor$. The required inequality $r \le \lfloor 3h/2 \rfloor + 1$ follows.

We now proceed to prove the 'if part' : for any given positives integers h and r that fulfill $h < r \le \lfloor 3h/2 \rfloor + 1$, we exhibit a segment space $(X(h,r), \mathcal{C}(h,r))$ with Helly number h and Radon number r. Keeping h fixed, and letting r run from $h+1$ to $\lfloor 3h/2 \rfloor + 1$, our space will be obtained as a convex subspace of the space $(X(h, \lfloor 3h/2 \rfloor + 1), \mathcal{C}(h, \lfloor 3h/2 \rfloor + 1))$, corresponding to the largest admissible value of r. This later space is referred to as $(X(h), \mathcal{C}(h))$ in the sequel.

The case $h = 1$.
Here, 2 is the only admissible value for the Radon number. Set $X(1) = \{a, b, c, d\}$ and $\mathcal{C}(1) = \{\emptyset, \{d\}, \{a, d\}, \{b, d\}, \{c, d\}, \{a, c, d\}, \{b, c, d\}, \{a, b, c, d\}\}$. This convexity is determined by the following non trivial convex dependencies: $d \in \langle a \rangle \cap \langle b \rangle \cap \langle c \rangle$ and $c \in \langle a, b \rangle$. Clearly, the Carathéodory number is 2 (for every 3-element subset is convexly dependent and $c \in \langle a, b \rangle \backslash (\langle a \rangle \cup \langle b \rangle)$), the Helly number is 1 (for d belongs to the core of every 2-element subset), and the Radon number is 2 (since d constitutes a partition-point for any 2-element subset). □

The even extremal case $h = 2m$ and $r = 3m + 1$, with $m \in \mathbb{N}_+$.
We set $X(2m) = \sqcup_{k=1,\ldots,m} \{t_k^1, t_k^2, t_k^3, \theta_k\}$, where the t_k^i's and the θ_k's are all distinct. We endow $X(2m)$ with a convexity $\mathcal{C}(2m)$ that satisfy the following property: for each $k \in \{1, \ldots, m\}$, the triple $T_k = \{t_k^1, t_k^2, t_k^3\}$ forms a triad for the set $S = \cup_k T_k$, with θ_k as associated triadic point. Explicitly, we put $\mathcal{C}(2m) = \{C \in 2^X; \mathrm{J}(C) \subseteq C\}$ where J is the binary operator defined by:

$$\mathrm{J}(t_k^i, t_k^j) = \theta_k \text{ for } k \in \{1, \ldots, m\},\ 1 \le i, j \le 3, i \ne j \qquad (5)$$

while $\mathrm{J}(x, y) = \emptyset$ for all other pairs $(x, y) \in X \times X$. Let us show that $\mathcal{C}(2m)$ has Carathéodory number 2, Helly number $2m$ and Radon number $3m+1$. In what follows, the notations $\langle \cdot \rangle$ and $\heartsuit$ refer to the convex hull operator relative to $\mathcal{C}(2m)$.

Claim 1. $\mathcal{C}(2m)$ has Carathéodory number 2. Indeed, it suffices to observe that any convex dependency of the form $a \in \langle A \rangle \backslash A$ implies that a is necessarily obtained by application of the operator J; so, for some index k, we have $a = \theta_k = \mathrm{J}(t_k^i, t_k^j)$ with $t_k^i, t_k^j \in \langle A \rangle \cap T_k$. Since J generates no element in T_k, we have $t_k^i, t_k^j \in A$, hence a is in the convex closure of a 2-element A-subset, as claimed.

The same reasoning leads also to the following statement:

Claim 2. For any $A \subseteq X$, $\langle A \rangle = A \cup \{\theta_k;\ |A \cap T_k| \geq 2\}$.

From Claim 2, it follows that the set S is Radon-independent for $S = A \sqcup B$ implies

$$\{\theta_k;\ |A \cap T_k| \geq 2\} \cap \{\theta_k;\ |B \cap T_k| \geq 2\} = \emptyset \tag{6}$$

(since no T_k has more than 3 points), hence $\langle A \rangle \cap \langle B \rangle = \emptyset$. Thus, the Radon number is at least $|S| + 1 = 3m + 1$.

Claim 3. $\mathscr{C}(h)$ has convex rank $3m$ and Radon number $3m + 1$. It suffices to remark that, by the pigeon-hole principle, any subset of $X(2m)$ with $3m + 1$ points contains some set of the form $T_k + \theta_k$. Such a set is convexly-dependent. The claim follows.

Claim 4. $\mathscr{C}(h)$ has Helly number $2m$. Let $A = \cup_{k=1,\ldots,m}\{t_k^1, t_k^2\}$. Claim 2 implies that for any index k and for any $a \in \{t_k^1, t_k^2\}$, we have $a, \theta_k \notin \langle A - a \rangle$. It follows that $\cap_{a \in A}\langle A - a \rangle = \emptyset$, i.e. A is Helly independent. Thus, the Helly number is at least $2m$. Let us prove now that any subset A of $2m + 1$ points in $X(2m)$ is Helly dependent. By the pigeon-hole principle, there exists some index k for which $|A \cap (T_k + \theta_k)| \geq 3$. If $\theta_k \in A$, then $|A \cap T_k| \geq 2$ hence $\theta_k \in \langle A - \theta_k \rangle$; so, A is convexly dependent and *a fortiori* Helly dependent. If $\theta_k \notin A$, then A contains T_k, which, clearly, is Helly dependent; hence A is Helly dependent too. Our claim follows.

The proof of the case $h = 2m$ and $r = 3m + 1$ is complete. □

The odd extremal case $h = 2m + 1$ and $r = 3m + 2$, with $m \in \mathbb{N}_+$.
Applying the above construction with $m+1$ instead of m, we obtain a segment convexity space $(X(2(m+1)), \mathscr{C}(2(m+1)))$ with Helly number $2(m+1)$ and Radon number $3(m+1)+1$. Following the same lines as in the preceding case, and keeping the same notations, it is straightforward to check that the convex subspace of $(X(2(m+1)), \mathscr{C}(2(m+1)))$ obtained by deletion of points t_{m+1}^2, t_{m+1}^3 and θ_{m+1} has Helly number $2m + 1$ and Radon number $3m + 2$, as required. This subspace is taken as $(X(2m+1), \mathscr{C}(2m+1))$. □

The general case.
Assume that h and r are integers such that $2 \leq h < r < \lfloor 3h/2 \rfloor + 1$. Set $h = 2m + \epsilon$ with $\epsilon \in \{0, 1\}$. Observing that $\lfloor 3h/2 \rfloor = 3m + \epsilon$, we may write r as $h + 1 + q$ where $q < m$. Referring to the notations introduced above for the extremal cases, we consider the space $(X(2m+\epsilon), \mathscr{C}(2m+\epsilon))$ and define

$$X(h, r) = \cup_{k=1,\ldots,q}\{t_k^1, t_k^2, t_k^3, \theta_i\} \cup \cup_{k=q+1,\ldots,m}\{t_k^1, t_k^2, \theta_k\} \cup \{t_{m+\epsilon}^1\} \tag{7}$$

Reasoning as above, the alert reader can easily check that $X(h, r)$ induces a convex subspace of $(X(2m+\epsilon), \mathscr{C}(2m+\epsilon))$ with Helly number h and Radon number r. □

The proof of theorem 5.2 is complete.

Corollary 5.3. *Positive integers h and r are respective Helly and Radon numbers of some segment space if and only if $h < r \leq \lfloor 3h/2 \rfloor + 1$.* □

Corollary 5.4. *Every integers h and r such that $2 \leq h < r \leq \lfloor 3h/2 \rfloor + 1$ are realizable respectively as Helly and Radon numbers of a finite geodesic convex geometry with Carathéodory number 2.*

Proof. Assuming $h > 1$, we prove that the convexity space $(X(h,r), \mathscr{C}(h,r))$, introduced in the above proof of theorem 5.2, possesses all the required properties. For all admissible values of h and r, the ground set is given by formula (7), where $h = 2m+\epsilon$ with $\epsilon \in \{0, 1\}$, and $r = h+1+q$ with $0 \leq q \leq m$. Note that the last point t^1_{m+1} is taken into account only in case $\epsilon = 1$. For $k = 1, \ldots, m+\epsilon$ we denote by X_k the set of all points with lower index k, and for $k = 1, \ldots, m$, we set $T'_k = X_k - \theta_k$; additionally, when $\epsilon = 1$, we put $T'_{m+1} = \{t^1_{m+1}\}$. Since $(X(h,r), \mathscr{C}(h,r))$ is a convex subspace of $(X(h, \lfloor 3h/2 \rfloor + 1), \mathscr{C}(\lfloor 3h/2 \rfloor + 1))$, visualizing formula (6), we see that the convex hull of a subset $A \subseteq X(h,r)$ is given by the following formula:

$$\langle A \rangle = A \cup \{\theta_k; |A \cap T'_k| \geq 2\} \tag{8}$$

To prove that $\mathscr{C}(h,r)$ is a *convex geometry*, we show that any convex set is the convex hull of its extreme points. What the extreme points of a convex set C look like? Setting $C' = C \cap \bigcup_{k=1,\ldots,m+\epsilon} T'_k$, we observe that any point $x \in C'$ is an extreme point of C since, by formula (8), $C - x$ is convex. Now let $\theta_k \in C \backslash C'$. If $|C \cap T'_k| \leq 1$, then, by (8), $C - \theta_k$ is convex, hence θ_k is an extreme point of C; otherwise, by (8) again, we have $\theta_k \in \langle C \cap T'_k \rangle \subseteq \langle C' \rangle$. In both eventualities, θ_k does belong to the convex hull of extreme points of C. To resume, any point of C belongs to the convex hull of the extreme points of C, as expected.

We now prove that $\mathscr{C}(h,r)$ is a *geodesic convexity*. We define $\mathrm{d}: X(h,r) \times X(h,r) \to \mathbb{N}$ as the only symmetric function that fulfills the following conditions:

(i) $\mathrm{d}(x,x) = 0$ for $x \in X$
(ii) $\mathrm{d}(x,y) = 4$ for $x, y \in T'_k$, $x \neq y$, where $1 \leq k \leq m$.
(iii) $\mathrm{d}(x,y) = 3$ for $x \in X_k$, $y \notin X_k$, where $1 \leq k \leq m+\epsilon$
(iv) $\mathrm{d}(x,\theta_k) = 2$, for $x \neq \theta_k$, where $1 \leq k \leq m$.

As readily seen, d is a distance on $X(h,r)$. In the metric space $(X(h,r), \mathrm{d})$ the only possible strict betweenness relations have the form $\mathrm{d}(x,y) = \mathrm{d}(x,\theta_k) + \mathrm{d}(\theta_k, y)$ with $x, y \in T'_k$ and $1 \leq k \leq m$. A comparison with formula (5) shows that the geodesic convexity of $(X(h,r), \mathrm{d})$ coincides with the convexity on $X(h,r)$ generated by the operator J, i.e. with convexity $\mathscr{C}(h,r)$. □

6. Comments and open questions

We showed in [9] how the Helly and Radon numbers of a general convexity space $\mathscr{C}$ can be realized by the geodesic convexity of some finite graph associated to $\mathscr{C}$. The generic graph construction presented in [9] (whose analysis and proof will appear in [10]) does not preserve the Carathéodory number: it is larger in the associated graph than in the original convexity. Nevertheless, when dealing with the particular convexity $\mathscr{C}(h,r)$ we introduced in section 5, it turns out that the generic construction can be modified in order to maintain the Carathéodory number to its original value 2. The geodesic convexity of the resulting graph is not any more a convex geometry. Details and proofs will appear in a forthcoming paper.

Another direction of further interest is the generalization of the present work to t-partition numbers (see [12]) for $t > 2$, with the particular motivation to test the partition conjecture (*ibid.*) in segment spaces: thanks to the 'triads', the question could be accessible.

What happens with larger Carathéodory numbers? As shown here, the Eckhoff-Jamison bound on the Radon number is not sharp for $c = 2$ and we got the impression that such a phenomenon is general (at least when $c \leq h$). We are still unable to emit a precise conjecture, and offer the following challenge.

Problem 6.1. In convexity spaces, can the ratio $\frac{r}{h \log c}$ become arbitrarily large?

References

[1] I. Barany and S. Onn, Carathéodorys theorem: colorful and applicable, in Bolyai Society Mathematical Studies (eds.) I. Barany and K. Borotczky, 6, North-Holland (1997) 11–21.

[2] C. Berge and M. Las Vergnas, Sur un théorème de type König pour hypergraphes, *Ann. New York Acad. Sci.*, **175** (1970) 32–40.

[3] C. Berge and P. Duchet, A generalisation of Gilmore's theorem, in Recent Advances in Graph Theory, Proc. 2nd Czekoslovack Sympos., Prague Academia, Prague (1975) 49–55.

[4] J. Calder, Some elementary properties of interval convexities, *J. London Math. Soc.*, **2, 3** (1971) 422–428.

[5] P. M. Cohn, Universal Algebra, Harper and Row, New York, 1965. New ed., D. REidel Pub. Co., Dortrecht-Boston-London (1981) 413 p.

[6] P. Duchet, Convexity in combinatorial structures, *Rend. Circ. Mat. Palermo (2) Suppl.*, **14** (1987) 261–293.

[7] P. Duchet, Convex sets in graphs II. Minimal path convexity, *J. Combin. Theory Ser. B.*, **44** (1988) 307–316.

[8] P. Duchet, Hypergraphs, Chapter 7 in Handbook of Combinatorics, (eds.) R. L. Graham, M. Grötschel and L. Lovász, 2 vol., North Holland Publ. Co., 1995 – Cambridge, MA: LIT Press (1996) 381–432.

[9] P. Duchet, Discrete convexity: retractions, morphisms and the partition problem. Proceedings of the conference on graph connections (Cochin, Jan. 28–31 1998), (eds.) R. Balakrishnan, H. M. Mulder and A. Vijayakumar, Allied Pub., New Delhi (1999) 10–18. MR# 2000k:05102.

[10] P. Duchet, Convex sets in graphs III: the geodetic convexity and the partition problem. (in preparation, to be submitted to European J. of Combinatorics).

[11] J. Eckhoff, Helly, Radon and Carathéodory type theorems, in Handbook of convex geometry (eds.) P. M. Gruber and J. Wills, Vol. A, B, North Holland, Amsterdam (1993) 389–448.

[12] J. Eckhoff, The partition conjecture, *Discrete Mathematics*, **221** (2000) no. 1, 61–78.

[13] P. H. Edelman and R. E. Jamison, The theory of convex geometries, *Geometriae Dedicata*, **19** (1985) 247–270.

[14] R. L. Graham, M. Grötschel and L. Lovász (eds.), Handbook of Combinatorics, 2 vol., Cambridge, MA: LIT Press (1996).

[15] F. Harary and J. Nieminen, Convexity in graphs, *J. Differ. Geom.*, **16** (1981) 185–190.

[16] R. E. Jamison-Waldner, A perspective on abstract convexity: classifying alignments by varieties, in *Convexity and Related Combinatorial Geometry*, Proc. 2nd Oklahoma Conf. (eds.) D. C. Kay and M. Breen, New York (1982) 113–150.

[17] R. E. Jamison-Waldner, Partition numbers for trees and ordered sets, *Pacific J. Math.*, **96** (1981) 115–140.

[18] D. C. Kay and E. W. Womble, Axiomatic Convexity Theory And Relationships Between The Carathéodory, Helly, and Radon Numbers, *Pacific J. Math.*, **38** (1971) no. 2, 471–485.

[19] F. W. Levi, On Helly's theorem and the axioms of convexity, *J. Indian Math. Soc.*, **15** (1951) 65–76.
[20] L. Lovász, Combinatorial problems and exercises, Akadémiai Kiadó/North-Holland, Budapest/Amsterdam (1979).
[21] S. T. S. Motzkin, Linear inequalities, mimeographed Lecture Notes, Univ. of California, Los Angeles (1951).
[22] G. Sierksma, Carathéodory and Helly numbers of convex-product-structures, *Pacific J. Math.*, **61** (1975) 275–282.
[23] G. Sierksma, Relationships between Carathéodory, Helly, Radon and exchange numbers of convexity spaces, *Nieuw Archief voor Wisk.*, **25** (1977) no. 3, 115–132.
[24] M. L. J. van de Vel, Theory of Convex Structures, North Holland, Amsterdam (1993).

Joint Proc. Int. Wrk.shps. Conv. in Discr. Struct. and Metric & Conv. Graph Th. No. 5, 2008, pp. 73–79.

On a metric related Transit Function and its Betweenness

Prasanth G. Narasimha-Shenoi

Department of Mathematics, Government College, Chittur, Palakkad 678 104, India
e-mail: gnprasanth@gmail.com

Abstract. The metric related transit function I_1 on a connected graph G is defined by $I_1(u, v)$ being the set of all vertices lying on all u, v-paths of length at most $d(u, v) + 1$ for any pair of vertices u, v of G, where d is the geodesic distance function of G. In this paper we study the transit function I_1 on G. We discuss the betweenness axioms for this function.

Keywords. Transit function, I_1-path, betweenness.

Subject Classification: 05C99, 05C12, 05C38, 05C75

1. Introduction

The term transit function specifically on a discrete set was introduced with the purpose of studying how to move around in discrete structures, like a graph, a poset or a similar structure, see [10]. There are several studies on transit functions or interval functions on graphs in the literature, see [1], [2], [4], [6], [7], [9]. A notable reference is [5] where several transit functions on graphs are surveyed in connection with their relation to intervals, convexities and betweenness. Well-known transit functions on graphs are defined in terms of paths in the graph, the idea of which is presented using what is known as a Φ-path in [5]. A Φ-path in a graph G is a path having property Φ, such as for example, 'geodetic', 'induced' etc. Formally, we take Φ to be a subfamily of the family of all paths in G. For $u, v \in V(G)$, then $\Phi(u, v)$ denotes the set of all paths in Φ between u and v. We call a path property a feasible path property if $\Phi(u, v) \neq \emptyset$, for all $u, v \in V$. Given a feasible path property Φ, the Φ-path transit function R_Φ on G is defined by $R_\Phi(u, v) = \{x \in V | x \text{ is on some } \Phi\text{-path in } G\}$. Examples of known path transit functions are the *geodesic*, the *induced* and the *all-paths* transit functions on G. Some references are: [6], [9], [10], [11], [12] for the geodesic transit function, [1], [2], [8] for the induced path transit function, and [3], [4], [13] for all-paths transit function. In [5] the transit function denoted by I_j is introduced, where $I_j(u, v)$ consists of all vertices on u, v-paths of length at most $d(u, v) + j$. In this paper our focus will be on the I_j-path transit function, in particular for $j = 1$. In Section 2 we define the transit function I_j and some definitions which are needed in the sequel. In Section 3, we discuss the properties of I_1 including betweenness.

2. Preliminaries

Throughout this paper $G = (V, E)$ denotes a finite, simple, loopless, connected graph with vertex set V and edge set E. A *transit function* on a set V is a function

$R : V \times V \to 2^V$ satisfying the following conditions

$$(t1) \quad u \in R(u, v) \text{ for all } u, v \in V(G)$$

$$(t2) \quad R(u, v) = R(v, u) \text{ for all } u, v \in V(G)$$

$$(t3) \quad R(u, u) = \{u\} \text{ for all } u \in V(G).$$

Examples of widely discussed transit functions are the interval function I on a graph G, the induced path function J, and the all-paths function A on a connected graph G. The interval function I is defined by

$$I(u, v) = \{w | w \text{ lies on a shortest } u, v\text{-path in } G\}$$

The transit functions J and A are defined as follows: $J(u, v)$ consists of the vertices on the induced u, v-paths, and $A(u, v)$ consists of the vertices on all u, v-paths in G.

Two transit functions derived from I and J are the triangle-path transit functions, namely I^{Δ} and J^{Δ} defined as follows:

$$I^{\Delta}(u, v) = \{w | w \text{ lies on a shortest } u, v\text{-path or is adjacent to two consecutive vertices on a shortest } u, v\text{-path }\},$$

$$J^{\Delta}(u, v) = \{w | w \text{ lies on an induced } u, v\text{-path in } G \text{ or adjacent to two consecutive vertices in an induced } u, v\text{-path in } G\}$$

The transit function I_j is defined as

$$I_j(u, v) = \{w | d(u, w) + d(w, v) \leq d(u, v) + j\}.$$

For a transit function R on V, the betweenness axioms were introduced by Morgana and Mulder in [8]. There are several notions of betweenness in the literature, see for example [3]. Let R be any transit function on a non-empty set V. The *betweenness axioms* on R are defined as follows.

$$(b1) \quad x \in R(u, v) \Rightarrow v \notin R(u, x)$$

$$(b2) \quad x \in R(u, v) \Rightarrow R(u, x) \subseteq R(u, v)$$

$$(b3) \quad x \in R(u, v), y \in R(u, x) \Rightarrow x \in R(y, v)$$

In this paper we slightly modify axiom $(b3)$ as follows

$$(b3') \quad x \in R(u, v), u, v \neq y \in R(u, x) \Rightarrow x \in R(y, v)$$

In [8] a *betweenness* is defined as a transit function satisfying axioms $(b1)$ and $(b2)$.

For a given graph G and another graph G' we say that G is G'-free if and only if G does not contain G' as an induced subgraph. In the literature a Θ-graph consists of three internally disjoint paths between two vertices. For our purposes we define these slightly different: a Θ-graph consists of a vertex u and an edge vx, with $d(u, v) = d(u, x)$, together with internally disjoint paths P, Q, and R such that P is a shortest u, v-path, Q is a shortest u, x-path, and R is a u, x-path of length $d(u, x) + 1$. Note that P, Q, and the edge vx form an odd cycle C of length $2d(u, v) + 1$. For the smallest two Θ-graphs, see the figure below, where R is denoted by dashed lines.

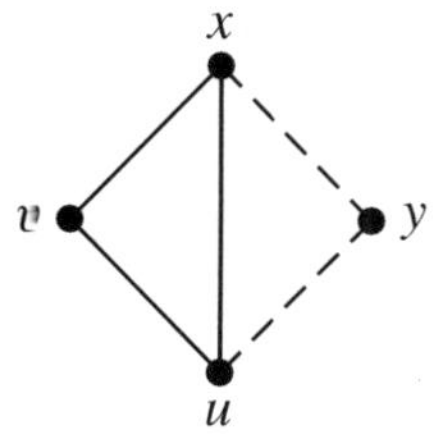

$K_4 - e$

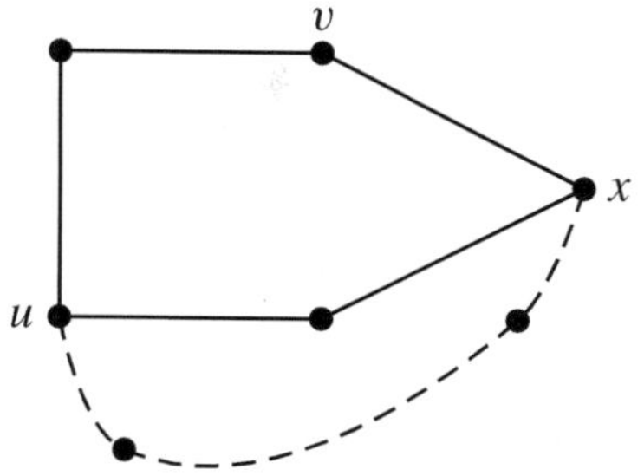

Θ-graphs

It is easy to see that the geodesic transit function I satisfies axioms $(b1)$, $(b2)$, and $(b3)$ on every graph. The induced path transit function J on the other hand need not always satisfy the betweenness axioms. The function J satisfies the betweenness axioms $(b1)$ and $(b2)$ if and only if G is *HHD*-free, and J satisfies the axioms $(b1)$, $(b2)$, and $(b3)$ if and only if G is distance hereditary, see [8]. Here *HHD* signifies the graphs *house* (a 5-cycle with one chord), *hole* (cycles of length at least 5), and *domino* (a 6-cycle with a chord between diametrical vertices). Similarly, for the monotone axiom the forbidden graphs are identified in [2]. It is easy to see that the transit function A satisfies $(b2)$ for any graph and satisfies $(b1)$ if and only if G is a tree [4].

By a triangle we mean the complete graph on three vertices. By long even house we mean an even cycle on which a triangle is attached as in the following figure. By $u \to y \to \cdots \to v$ we mean that it is a u, v-path in which u and y are adjacent whereas y and v are not adjacent.

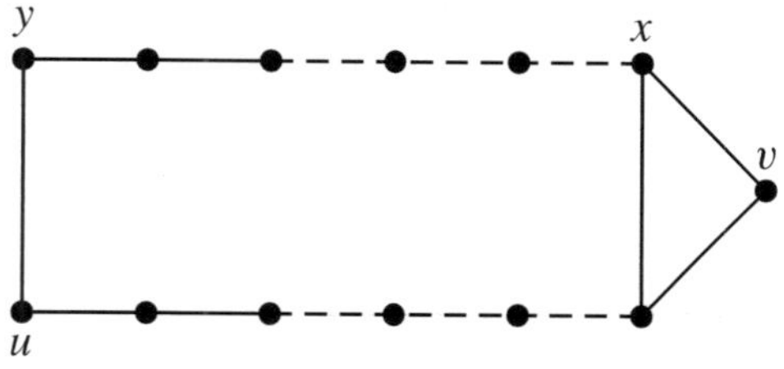

Long even house

For both I^{Δ} and I_j, if G is a K_3, say on u, v, w, then we see that axiom $(b3)$ fails. Since $w \in R(u, v)$, and $v \in R(u, w)$ but $w \notin R(v, v)$, for $R = I^{\Delta}$ or I_j. Because of this reason we consider axiom $(b3')$ instead of axiom $(b3)$.

3. The I_1-path transit function

In this section we study the properties of the path transit function I_1. We have the following observations.

Observation 1. $I^{\Delta}(u, v) \subseteq I_1(u, v)$, *for all* $u, v \in V(G)$.

Observation 2. *Let G be a connected graph, and let $j \geq 2diam(G)$. Then $I_j(u, v) = V(G)$, for all distinct u, v in G.*

Observation 3. *If G is a complete graph, then $I \neq I^{\Delta} = I_1 = I_2 = \cdots$.*

Observation 4. *For all j, $I_j(u, v) \subseteq I_{j+1}(u, v)$.*

Observation 5. *If $x \in I_1(u, v)$ and $uv \in E(G)$ then $ux, xv \in E(G)$.*

Lemma 1. *$I(u, x) \subseteq I_1(u, v)$, for all $x \in I_1(u, v)$.*

Proof. Let $x \in I_1(u, v)$. We distinguish two cases.

Case 1. $x \in I(u, v)$.
Then we have $I(u, x) \subseteq I(u, v) \subseteq I_1(u, v)$.

Case 2. $x \notin I(u, v)$.
Then we have $d(u, x) + d(x, v) = d(u, v) + 1$.
Let $y \in I(u, x)$. Then $d(u, y) + d(y, x) = d(u, x) \leq d(u, v) + 1$. Therefore $y \in I_1(u, v)$. Hence $I(u, x) \subseteq I_1(u, v)$. □

Lemma 2. *$I_1(u, v) = V(G)$ for all distinct $u, v \in V(G)$ if and only if G is complete.*

Proof. Let G be complete. Then $I_1(u, v) = V(G)$, for all distinct u,$v \in V(G)$.

Conversely, let $I_1(u, v) = V(G)$, for all distinct u, v in V. Suppose G is not complete. Then G contains at least two vertices u and x with $d(u, x) = 2$. Let v be a common neighbour of u and x. Now $d(u, v) = 1$ and there is no u, v-path of length 2 containing x. Hence $v \notin I_1(u, x)$, whence $I_1(u, x) \neq V(G)$. □

Lemma 3. *$I_1 = I$ if and only if G is bipartite.*

Proof. First let $I_1 = I$. Assume that G contains an odd cycle. We choose C to be a shortest odd cycle, so that C is isometric (that is, distances measured in C are equal to those measured in G). Then we can take two vertices $u, v \in C$ so that there exist two paths P_1 and P_2, where $P_1 = u \to \cdots \to v$ is of length $d(u, v)$, and P_2 is of length $d(u, v) + 1$ so that $u \to P_1 \to v \to P_2 \to u$ is a cycle of length $2d(u, v) + 1$, so that $I_1(u, v)$ contains C, in particular the path P_2. But I will not contain the path P_2 so that $I_1 \neq I$, a contradiction to the assumption.

Conversely, let G be bipartite. We have to show that $I_1 = I$. Assume to the contrary that $I_1 \neq I$. Then for some $u, v \in V(G)$, we have $I_1(u, v) \neq I(u, v)$. Then $I_1(u, v)$ will contain at least two u, v-paths of length $d(u, v)$ and $d(u, v) + 1$, respectively, so that these two paths together form a closed walk of length $2d(u, v) + 1$, which contains an odd cycle, a contradiction. □

Lemma 4. *I_1 satisfies axiom $(b1)$ if and only if G is bipartite.*

Proof. If G is bipartite, then, by the previous lemma, we have $I_1 = I$. Since I satisfies axiom $(b1)$, it follows that I_1 also satisfies axiom $(b1)$.

Conversely, let I_1 satisfy axiom $(b1)$. Then $x \in I_1(u, v)$, whence $v \neq I_1(u, x)$. Assume that G contains an odd cycle, and let C be a shortest odd cycle, so that C is isometric. In C we can find u, v such that there exists a vertex x with $d(u, x) = d(u, v)$ and $xv \in E(G)$. Now $x \in I_1(u, v)$ and $v \in I_1(u, x)$, which contradicts axiom $(b1)$. □

Lemma 5. *The path transit function I_j satisfies axiom $(b1)$, for $j \geq 2$, if and only if G is a tree.*

Proof. If G is a tree, then $I_j = I$, so I_j trivially satisfies axiom $(b1)$. So let G contains a cycle. We choose a shortest cycle C in G, so that C is isometric. Choose vertices u and v on C such that $d(u, v)$ is as large as possible, and let x be a neighbor of v on C. Then $d(u, x) \geq d(u, v) - 1$. So the u, x-path on C containing v is of length at most $d(u, x) + 2$. Hence we have that $v \in I_j(u, x)$, which contradicts axiom $(b2)$. □

Theorem 6. *The path transit function I_1 satisfies axiom $(b2)$ if and only if G is a Θ-free graph.*

Proof. First let G be Θ-free. Assume that I_1 does not satisfy axiom $(b2)$. Then, for some $u, v \in V(G)$ and $x \in I_1(u, v)$, we have $I_1(u, x) \nsubseteq I_1(u, v)$. Therefore there exists some $y \in I_1(u, x)$ with $y \notin I_1(u, v)$. If $x \in I(u, v)$, then, since $I(u, x) \subseteq I(u, v)$, we have $I_1(u, x) \subseteq I_1(u, v)$. Hence we have $x \in I_1(u, v) \setminus I(u, v)$. Therefore the cycle $u \to \cdots \to x \to \cdots \to v \cdots \to u$ is an odd cycle. Since $I(u, x) \subseteq I_1(u, v)$, we have $y \notin I(u, x)$, that is, there is a path $P_1 = u \to \cdots \to y \to \cdots \to x$ of length $d(u, x) + 1$. This path P_1 together with the odd cycle $u \to \cdots \to x \to \cdots \to v \to \cdots \to u$ is a Θ-graph. This is impossible.

Conversely, let I_1 satisfy axiom $(b2)$. Assume that G contains an induced Θ-subgraph. Then, for some $u, v \in V(\Theta)$, there exists a vertex x on a u, v-path P_1 of length $d(u, v) + 1$, and there exists a u, x-path P_2 of length $d(u, x) + 1$. Since $d(u, x) + d(x, v) = d(u, v) + 1$ we have that $l(P_1) + l(P_2) = d(u, x) + 1 + d(u, v) - d(u, x) + 1 = d(u, v) + 2$, where $l(P_1)$ is the length of the path P_1. Therefore there exists a vertex $y \in I_1(u, x)$ with $y \notin I_1(u, v)$, so that axiom $(b2)$ does not hold. Hence G is Θ-free. □

Corollary 7. *I_1 satisfies betweenness if and only if G is bipartite.*

Before we move on to prove axiom $(b3')$, we have the following observation.

Observation 8. *For axiom $(b3')$, if $uv \in E(G)$, and $x \in I^{\Delta}(u, v)$, $y \in I^{\Delta}(u, x)$, then we see that for $u, v \neq y$, $x \in I^{\Delta}(y, v)$. So for any two adjacent vertices u, v we see that axiom $(b3')$ is satisfied.*

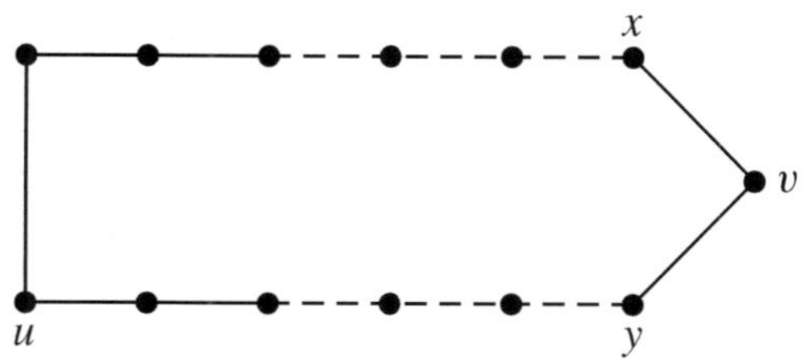

Odd Hole

Theorem 9. *I_1 satisfies axiom $(b3')$ if and only if G is paw-free and does not contain odd holes.*

Proof. First let G contain an induced odd cycle of length at least 5, and let C be a shortest such cycle. Then C is isometric. Choose the vertices $\{u, v, x, y\}$ as in the figure of the odd cycle. Then it is easy to check that for these vertices axiom $(b3')$ does not hold. Second, let G contain an induced paw. Choose the vertices $\{u, v, x, y\}$ as in the figure of the paw. Again it is easy to check that for these vertices axiom $(b3')$ does not hold.

Conversely, let G be paw-free and without odd holes. Assume that I_1 does not satisfy axiom $(b3')$. Then, for some $u, v \in V(G)$ and $x \in I_1(u, v)$, $u, v \neq y \in I_1(u, x)$, we have $x \notin I_1(y, v)$. We consider two cases.

Case 1. $x \in I(u, v)$.

If we would have $y \in I(u, x)$, then we have $x \in I(y, v)$. So $y \in I_1(u, x) \setminus I(u, x)$.

Suppose that y is adjacent to two consecutive vertices x_1, x_2 on some shortest u, x-path P. If $x_2 \neq x$, then let x_3 be the neighbour of x_2 on P closer to x. Since y is not in $I(u, x)$, it follows that y is not adjacent to x_3. But then y, x_1, x_2, x_3 induce a paw in G, which is forbidden. So we must have $x_2 = x$. Let x' be a neighbour of x on a shortest x, v-path P'. To avoid a paw, it follows that y is adjacent to x'. But now $u \rightarrow \cdots P \cdots \rightarrow x_1 \rightarrow y \rightarrow x' \rightarrow \cdots P' \cdots \rightarrow v$ is a path of length $d(u, x) + d(x, v)$, whence it must be a shortest u, v-path. Hence the path $y \rightarrow x \rightarrow \cdots P' \cdots \rightarrow v$ is a path of length $d(y, v) + 1$. This implies that $x \in I_1(y, v)$, which is impossible.

So y is *not* adjacent to two consecutive vertices on any shortest u, x-path. Let Q be a u, x-path of length $d(u, x) + 1$ containing y. Now we choose a shortest u, x-path P. Let x_1 be the last common vertex of Q and P before y and let x_2 be the first common vertex of Q and P after y. We may choose P such that $d(x_1, x_2)$ is as small as possible. Let P' and Q' be the subpaths between x_1 and x_2 of P and Q, respectively. Then P' is of length $d(x_1, x_2)$ and Q' is of length $d(x_1, x_2) + 1$. So $Q' \rightarrow P'$ is a cycle of length $2d(x_1, x_2) + 1 \geq 5$. Hence it can not be induced, and there is a chord $y'x'$ between an internal vertex y' of Q' and an internal vertex x' of P'. Assume that y' is between y and x_1 (otherwise replace x_1 by x_2 in the following argument). We may choose the chord such that y_1 is the vertex on Q' closest to x_1 with a chord to P', and amongst these chords we may choose the one with x' closest to x_1. Let P'' be the subpath of P between x_1 and x', and Q'' that of Q between x_1 and y'. Because of the choice of P with $d(x_1, x_2)$ as small as possible, the subpaths P'' and Q'' must have the same length. But then together with the chord $y'x'$ they form an induced odd cycle. To avoid an odd hole the length of P'' and Q'' must be 1, so x_1, x', y' form a triangle. Now y' can not be adjacent to the neigbour x'' of x' on P closer to x (for otherwise we get a conflict with the choice of P). So the triangle together with x'' induces a paw. This impossibility settles Case 1.

Case 2. $x \notin I(u, v)$.

Then there exists a u, v-path R of length $d(u, v) + 1$ containing x. Moreover, the subpath P of R between u and x is a shortest u, x-path. Note that we may take any shortest u, x-path to be P. Now, if y would be in $I(u, x)$, then it is easy to see that $x \in I^{\Delta}(y, v)$. So we have $y \notin I_1(u, x)$. So y is not on P, but there is a u, x-path Q of length $d(u, x) + 1$ containing y.

First suppose that y is adjacent to two consecutive vertices x_1, x_2 on P. If we would have $x_2 \neq x$, then let x_3 be the neighbour of x_2 on P closer to x. Then y, x_1, x_2, x_3 would induce a paw. So we must have $x_2 = x$. If we would have $u \neq x_1$, then let x_4 be the neighbour of x_1 on P closer to u. Then y, x_2, x_1, x_4 would induce a paw. So we have that $u = x_1$. Moreover, $d(u, v) = d(x, v)$. Now $y \to x \to \cdots R \cdots \to v$ is a path of length $d(u, x)+1$. So, to avoid that x is in $I_1(y, v)$ we must have that $d(y, v) < d(u, x)$. Since y and x are adjacent, it follows that $d(y, v) = d(x, v) - 1$. Since $y \neq v$, there is a neighbour y_1 of y with $d(y_1, v) = d(y, v) - 1 = d(x, v) - 2 = d(u, v) - 2$. But now u, x, y, y_1 induce a paw, which is forbidden.

So y is not adjacent to two consecutive vertices on P. Note that in Case 1 when we dealt with the subcase that y was not adjacent to two consecutive vertices on P we only used the shortest u, x-path P and a path of length $d(u, x) + 1$ containing y. Therefore, in a similar way as in this subcase of Case 1, we get to a contradiction. This settles Case 2, and the theorem. □

References

[1] M. Changat and J. Mathew, On triangle path convexity in graphs, *Discrete Math.*, **206** (1999) 91–95.

[2] M. Changat and J. Mathew, Induced path transit function, monotone and Peano axioms, *Discrete Math.*, **286** (2004) 185–194.

[3] M. Changat, J. Mathew and H. M. Mulder, Induced path function, Monotonicity and Betweenness (submitted).

[4] M. Changat, S.Klavžar and H. M. Mulder, The all-paths transit function of a graph, *Czech. Math. J.*, **51** (2001) no. 126, 439–448.

[5] M. Changat, H. M. Mulder and G. Sierksma, Convexities related to path properties on graphs, *Discrete Math.*, **290** (2005) 117–131.

[6] P. Duchet, Convexity in combinatorial structures. *Rend. Circ. Mat. Palermo (2) Suppl.*, **14** (1987) 261–293.

[7] P. Duchet, Convex sets in Graphs II Minimal path convexity, *J. Combin. Theory Ser B*, **44** (1988) 307–316.

[8] M. A. Morgana and H. M. Mulder, The induced path convexity, betweenness and svelte graphs, *Discrete Math.*, **254** (2002) 349–370.

[9] H. M. Mulder, The Interval Function of a Graph. Mathematical Centre Tracts 132, Mathematisch Centrum, Amsterdam (1980).

[10] H. M. Mulder, Transit functions on graphs (and posets), (in this volume) 117–130.

[11] L. Nebeský, A characterization of the interval function of a connected graph, *Czech. Math. J.* **44** (1994) 173–178.

[12] L. Nebeský, Characterzing the interval function of a connected graph, *Mathematica Bohemica*, **123** (1994) 137–144.

[13] E. Sampathkumar, Convex sets in graphs, *Indian J. Pure Appl. Math.*, **15** (1984) 1065–1071.

Joint Proc. Int. Wrk.shps. Conv. in Discr. Struct. and Metric & Conv. Graph Th. No. 5, 2008, pp. 81–85.

Fault-Tolerant Metric Dimension of Graphs*

Carmen Hernando[1], Merce Mora[2], Peter J. Slater[3] and David R. Wood[2]

[1]Department de Matemàtica Aplicada I, Universitat Politècnica de Catalunya, Barcelona, Spain
[2]Department de Matemàtica Aplicada II, Universitat Politècnica de Catalunya, Barcelona, Spain
[3]Mathematical Sciences Dept. and Computer Sciences Dept. University of Alabama in Huntsville, Huntsville, AL, USA 35899
e-mail: carmen.hernando@upc.edu; merce.mora@upc.edu; slater@math.uah.edu; pslater@cs.uah.edu; david.wood@upc.edu

Abstract. An ordered set S of vertices in a graph G is said to *resolve* G if every vertex in G is uniquely determined by its vector of distances to the vertices in S. The *metric dimension* of G is the minimum cardinality of a resolving set of G. In this paper we introduce the study of the fault-tolerant metric dimension of a graph. A resolving set S for G is *fault-tolerant* if $S \setminus \{v\}$ is also a resolving set, for each v in S, and the *fault-tolerant metric dimension* of G is the minimum cardinality of such a set. In this paper we characterize the fault-tolerant resolving sets in a tree T. We show that the fault-tolerant metric dimension values are bounded by a function of the metric dimension values independent of any graphs.

Keywords. Graph, distance, metric dimension.

Subject Classification: 05C12 (distance in graphs)

1. Introduction

For a graph G with vertex set $V(G)$ and edge set $E(G)$, the distance between two vertices u and v in $V(G)$ is the minimum number of edges in a $u-v$ path and is denoted by $d_G(u,v)$ or simply $d(u,v)$ if the graph G is clear. A vertex x *resolves* two vertices u and v if $d(x,u) \neq d(x,v)$. A vertex set $S \subseteq V(G)$ is said to be *resolving* for G if for every two distinct vertices u and v in $V(G)$ there is a vertex x in S that resolves u and v. The minimum cardinality of a resolving set of G is called the *metric dimension* of G and is denoted by $\beta(G)$. A resolving set of order $\beta(G)$ is called a *metric basis* of G.

Equivalently, for an ordered subset $S = (v_1, v_2, \ldots, v_k)$ of vertices in $V(G)$ the S-coordinates of a vertex x in $V(G)$ are $f_S(x) = (d(x,v_1), d(x,v_2), \ldots, d(x,v_k))$. Then S is a resolving set if for every two vertices x and y in $V(G)$ we have $f_S(x) \neq f_S(y)$.

These concepts were introduced for general graphs independently by Slater [6] and by Harary and Melter [4]. Resolving sets have since been widely investigated. (See the bibliographies of [1] and [3].)

*Research supported by the projects MCYT-FEDER-BFM2003-00368, Gen-Cat-2001SGR00224, MCYT-HU2002-0010, MEC MTM2006-01267, DURSI 2005SGR00692, and MCYT-FEDER BFM2003-00368. The research of David Wood is supported by a Marie Curie Fellowship from the European Commission under contract MEIF-CT-2006-023865.

As described in Slater [6], each v_i in S can be thought of as the site for a sonar or loran station, and each vertex location must be uniquely determined by its distances to the sites in S. In this paper we consider (single) fault-tolerant resolving sets S for which the failure of any single station at vertex location v in S leaves us with a set that still is a resolving set.

A resolving set S for a graph G is *fault-tolerant* if $S \setminus \{v\}$ is also resolving for each v in S. The *fault-tolerant metric dimension* of G is the minimum cardinality of a fault-tolerant resolving set, and it will be denoted by $\beta'(G)$. A fault-tolerant resolving set of order $\beta'(G)$ is called a *fault-tolerant metric basis*.

We consider fault-tolerant resolving sets for trees in section 2.. In section 3. we show that the fault-tolerant metric dimension values are bounded by a function of the metric dimension values independent of the given graph.

2. Fault-tolerant resolving sets for trees

Consider the tree T' in Figure 1. It will be seen below that $\beta(T') = 10$ and that $S = (1, 2, \ldots, 10)$ is a metric basis. We have, for example, that

$$f_S(x) = (11, 11, 11, 11, 10, 10, 10, 10, 1, 4),$$

$$f_S(y) = (11, 11, 11, 11, 10, 10, 10, 10, 3, 4),$$

$$f_S(v) = (8, 8, 8, 8, 3, 3, 7, 7, 8, 9),$$

$$f_S(t) = (8, 8, 8, 8, 1, 3, 7, 7, 8, 9).$$

Note that $f_S(x)$ and $f_S(y)$ agree everywhere except for the ninth component, that is, the only vertex in S that resolves vertices x and y is vertex 9. Likewise, only vertex 5 resolves vertices v and t. We will see that $\beta'(T') = 14$ and that $S \cup \{y, v, r, s\}$ is one example of a fault-tolerant metric basis of T'.

It is easy to see that for the path P_n on $n \geq 2$ vertices we have $\beta(P_n) = 1$ and $\beta'(P_n) = 2$. Observe that the unique fault-tolerant metric basis of P_n is formed by the two endpoints.

Henceforth we will only be concerned with trees T that have maximum degree $\Delta(T) \geq 3$. The degree of vertex v is denoted by $\deg(v)$.

A *branch* of a tree T at a vertex v is the subgraph induced by v and one of the components of $T \setminus \{v\}$. Note that each $v \in V(T)$ has $\deg(v)$ branches. A branch B of T at v which is a path will be called a *branch path* when $\deg(v) \geq 3$. The tree T' (see Figure 1) has sixteen branch paths. The vertex v in a branch path with $\deg(v) \geq 3$ will be called a *stem* of the branch path. Tree T' has six stems: $s_1, \ldots, s_6$. We let $L_1, \ldots, L_k$ be the components of the subtree induced by the set of all branch paths. Thus k is the number of stems, and each L_i can be obtained by subdividing edges starting with a star. For T' we have $k = 6$ and subdivisions: one $K_{1,5}$, three $K_{1,3}$ and two $K_{1,1}$.

Theorem 2.1 ([6]). *Let T be a tree of order $n \geq 3$. Vertex set S is a resolving set if and only if for each vertex u there are vertices from S on at least* $\deg(u) - 1$ *of the* $\deg(u)$ *components of $T \setminus \{u\}$.*

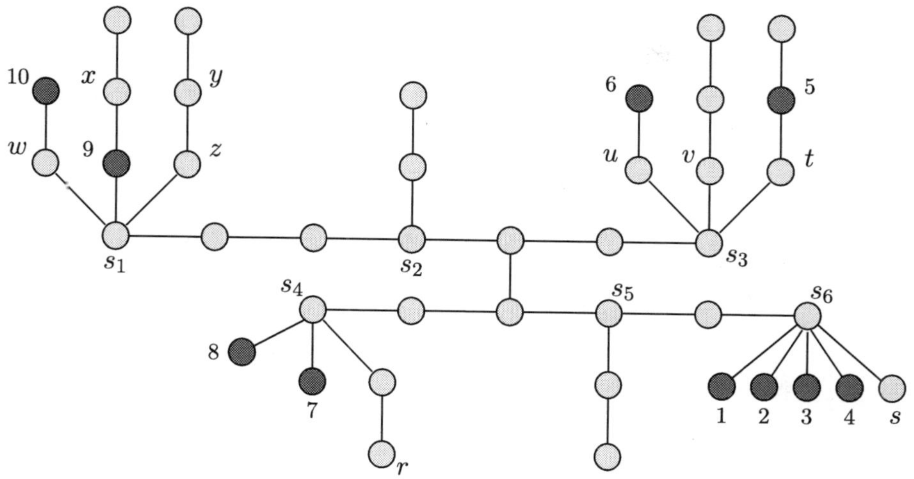

Figure 1. Tree T' with $\beta(T') = 10$ and $\beta'(T') = 14$.

Theorem 2.2 ([6]). *Let T be a tree with set L of endpoints with $|L| \geq 3$. Let $L_1, \ldots, L_k$ be the components of the subtree induced by the set of all branch paths, and let e_i be the number of branch paths in T that are in L_i. Then $\beta(T) = |L| - k$, and S is a metric basis if and only if it consists of exactly one vertex from each of exactly $e_i - 1$ of the branch paths of L_i, for each L_i, $1 \leq i \leq k$.*

Assume that S is a fault-tolerant resolving set for tree T. For each v in S, the set $S \backslash \{v\}$ is a resolving set. By Theorem 2.1, for any $v \in S \cap L_i$, the set $S \backslash \{v\}$ must contain a vertex from each of $e_i - 1$ of the branch paths of L_i. Thus, $|S \cap Li| \geq e_i$ when $e_i \geq 2$. If L is the set of endpoints and E_1 is the set of endpoints to branch paths where $e_i = 1$, then, by Theorem 2.1, $L \backslash E_1$ is a fault-tolerant resolving set. This implies the next theorem.

Theorem 2.3. *Let T be a tree with set L of endpoints with $|L| \geq 3$. Let $L_1, \ldots, L_k$ be the components of the subtree induced by the set of all branch paths, and let e_i be the number of branch paths in T that are in L_i. Let E_1 be the set of endpoints corresponding to branch paths where $e_i = 1$. Then $\beta'(T) = |L \backslash E_1|$ and $L \backslash E_1$ is a fault-tolerant metric basis.*

Remark 2.4. If S is a fault-tolerant metric basis for T and s_i is the stem vertex of L_i, then $e_i \geq 3$ implies that $S \cap L_i$ will consist of exactly one vertex from each path of $L_i \backslash \{s_i\}$, and, if $e_i = 2$, then $S \cap L_i$ can be any two vertices of $L_i \backslash \{s_i\}$.

In particular, for the tree T' of Figure 1, $\beta'(T') = 14$ and $(1, 2, 3, \ldots, 10, y, v, r, s)$ is one example of a fault-tolerant metric basis.

3. Relation between $\beta'(G)$ and $\beta(G)$

In this section we prove that fault-tolerant metric dimension is bounded by a function of metric dimension (independent of the graph). As usual, $N(v)$ and $N[v]$ denote vertex v's open and closed neighborhoods, respectively.

Lemma 3.1. *Let S be a resolving set of G. For each vertex $v \in S$, let $T(v) := \{x \in V(G) : N(v) \subseteq N(x)\}$. Then $S' := \cup_{v \in S}(N[v] \cup T(v))$ is a fault-tolerant resolving set of G.*

Proof. Consider a vertex $v \in S'$. If $v \notin S$ then $S' \setminus \{v\}$ resolves G since $S \subseteq S' \setminus \{v\}$. Now assume that $v \in S$.

Let p and q be distinct vertices of G. We must show that p and q are resolved by some vertex in $S' \setminus \{v\}$. If not, then v must resolve p and q since S resolves p and q. Without loss of generality $d(v, p) \leq d(v, q) - 1$.

First suppose that $p \neq v$. Let w be the neighbour of v on a shortest path between v and p. Then $w \in S' \setminus \{v\}$ and $d(v, p) = d(w, p) + 1$. Thus $d(w, p) + 1 \leq d(v, q) - 1$. Now $d(v, q) - 1 \leq d(w, q)$. Hence $d(w, p) + 1 \leq d(w, q)$. Thus w resolves p and q.

Now assume that $p = v$. If $q \in S'$ then $q \in S' \setminus \{v\}$ and q resolves p and q. Otherwise $d(v, q) \geq 2$ and q is not adjacent to some neighbour w of v. Then $d(v, w) = 1$ and $d(q, w) \geq 2$. Thus $w \in S' \setminus \{v\}$ resolves v $(= p)$ and q.

Hence S' is a fault-tolerant resolving set of G. □

The following lemma is implicit in [2,5]. We include the proof for completeness.

Lemma 3.2. *Let S be a resolving set in a graph G. Then for each vertex $v \in S$, the number of vertices of G at distance at most k from v is at most $1 + k(2k + 1)^{|S|-1}$.*

Proof. Say $1 \leq d(v, w) \leq k$. For every vertex $u \in S$ with $u \neq v$, we have $|d(w, u) - d(w, v)| \leq k$. Thus there are $2k + 1$ possible values for $d(w, u)$, and there are at most k possible values for $d(w, v)$. Thus the vector of distances from w to S has $k(2k + 1)^{|S|-1}$ possible values. The result follows, since the vertices at distance at most k are resolved by S. □

Theorem 3.3. *Fault-tolerant metric dimension is bounded by a function of the metric dimension (independent of the graph). In particular, $\beta'(G) \leq \beta(G)(1 + 2 \cdot 5^{\beta(G)-1})$ for every graph G.*

Proof. Let S be a metric basis for a graph G. Lemma 3.2 with $k = 2$ implies that $|N[v] \cup T(v)| \leq 1 + 2 \cdot 5^{\beta(G)-1}$ for each vertex $v \in S$. Thus $|S'| \leq \beta(G)(1 + 2 \cdot 5^{\beta(G)-1})$. □

References

[1] José Cáceres, Carmen Hernando, Mercè Mora, Ignacio M. Pelayo, María L. Puertas, Carlos Seara and David R. Wood, On the metric dimension of cartesian products of graphs, *SIAM J. Discrete Math.*, **21** (2007) no. 2, 423–441.

[2] Gary Chartrand, Linda Eroh, Mark A. Johnson and Ortrud R. Oellermann, Resolvability in graphs and the metric dimension of a graph, *Discrete Appl. Math.*, **105** (2000) no. 1–3, 99–113.

[3] Gary Chartrand and Ping Zhang, The theory and applications of resolvability in graphs, A survey, Proc. 34th Southeastern International Conf. on Combinatorics, Graph Theory and Computing, vol. 160 of *Congr. Numer.*, (2003) 47–68.

[4] Frank Harary and Robert A. Melter, On the metric dimension of a graph, *Ars Combinatoria*, **2** (1976) 191–195.

[5] Samir Khuller, Balaji Raghavachari and Azriel Rosenfeld, Landmarks in graphs, *Discrete Appl. Math.*, 70 (1996) no. 3, 217–229.

[6] Peter J. Slater, Leaves of trees, Proc. 6th Southeastern Conf. on Combinatorics, Graph Theory, and Computing, vol. 14 of *Congressus Numerantium* (1975) 549–559.

Joint Proc. Int. Wrk.shps. Conv. in Discr. Struct. and Metric & Conv. Graph Th. No. 5, 2008, pp. 87–95.

Hunting for Cubic Partial Cubes

Sandi Klavžar

Department of Mathematics and Computer Science. FNM, University of Maribor, Koroška cesta 160. 2000 Maribor, Slovenia
and
Institute of Mathematics, Physics and Mechanics, Jadranska 19, 1000 Ljubljana
e-mail: sandi.klavzar@uni-mb.si

Abstract. Partial cubes are graphs that admit isometric embeddings into hypercubes. Different methods and constructions that lead to cubic partial cubes are described. A further direction for the search of cubic partial cube is indicated and it is asked whether the Desargues graph is the only cubic partial cube that is not planar.

Keywords. Partial cube, cubic graph, Cartesian product of graphs, simplicial arrangement.

2000 Subject Classification: 05-02, 05C12, 05C75

1. Introduction

Isometric subgraphs of hypercubes are nowadays known as partial cubes. Among prominent examples of partial cubes let us mention trees, even cycles, benzenoid graphs [13], and median graphs [18]. Moreover, given partial cubes G and H, their Cartesian product $G \,\square\, H$ is a partial cube again.

Partial cubes possess many appealing structural properties, and have also found numerous applications, for instance in routing [11], oriented matroids [10], chemical graph theory [8], and psychology [20].

For one of the most important subclasses of partial cubes, median graphs, Mulder [17] proved that hypercubes are the only regular median graphs. In other words, the only regular median graphs are Cartesian products of several copies of K_2. In this paper we survey the state of the art of the following problem:

Problem 1.1. Classify cubic partial cubes.

Even prisms form trivial examples of cubic partial cubes. In Section 3. we consider several approaches that were used in the search for nontrivial cubic partial cubes. It is noted that among tree-like partial cubes there are no such examples and that among the generalized Petersen graphs only the Desargues graph is such. Cubic inflation is also introduced and a list of 15 non-trivial examples listed. In the subsequent section the concept of a tribe is briefly explained. The concept leads to many new examples. Eppstein's ideas from [9] are presented in Section 5., leading to several infinite families

*Supported in part by the Ministry of Science of Slovenia under the grant P1-0297.

and many sporadic examples of the quested graphs. We conclude by proposing a combination of the ideas of this paper to get as complete list of cubic partial cubes as possible. We also ask whether the Desargues graph is the only cubic partial cube that is not planar.

2. Definitions and expansions

The distance considered in this paper is the usual shortest path distance d. A subgraph H of a connected graph G is *convex* if for any vertices u, v of H any shortest path between u and v in G lies completely in H. Relaxing the condition "any shortest path" to "at least one shortest path", the isometry concept appears. More precisely, a subgraph H of a connected graph G is an *isometric subgraph* if $d_H(u, v) = d_G(u, v)$ holds for any vertices $u, v \in H$.

The *hypercube* or *n-cube* Q_n, $n \geq 1$, is the graph with vertex set $\{0, 1\}^n$, two vertices being adjacent if the corresponding tuples differ in precisely one place. Then $Q_1 = K_2$, $Q_2 = C_4$, and Q_3 is the skeleton of the usual solid 3-dimensional cube.

A graph G is a *partial cube* if G is an isometric subgraph of some n-cube. A graph G is a *median graph* if there exists a unique vertex x to every triple of vertices u, v, and w such that x lies simultaneously on a shortest u, v-path, shortest u, w-path, and shortest w, v-path. It is well-known that median graphs are partial cubes.

The *Cartesian product* $G \,\Box\, H$ of two graphs has the vertex set $V(G) \times V(H)$ where the vertex (g, h) is adjacent to (g', h') whenever $gg' \in E(G)$ and $h = h'$, or $g = g'$ and $hh' \in E(H)$. An equivalent description of the n-cube Q_n is that Q_n is the Cartesian product of n copies of the complete graph on two vertices K_2.

In the rest of this preliminary section we describe the concept of an expansion which is the most fundamental (nontrivial) concept in the theory of partial cubes. It is due to Mulder [16,17] who introduced it to characterize median graphs, while 10 years later Chepoi [4] realized that the concept is useful also in the more general framework of partial cubes (and, in fact, also more generally for partial Hamming graphs).

Let G be a connected graph. Then two nonempty isometric subgraphs G_0 and G_1 form an *isometric cover* of G provided that $G = G_0 \cup G_1$, by which we mean that $V(G) = V(G_0) \cup V(G_1)$ and $E(G) = E(G_0) \cup E(G_1)$. For $i = 0, 1$, let G_i' be an isomorphic copy of G_i, and for a vertex $u \in G_0 \cap G_1$, let u_i be the corresponding vertex in G_i'. The *expansion* of G with respect to G_0, G_1 is the graph G' obtained from the disjoint union of G_0' and G_1', where for each $u \in G_0 \cap G_1$ the vertices u_0 and u_1 are joined by a new edge in G'. It is demonstrated in Fig. 1 that C_8 is an expansion of C_6.

The expansion of G with respect to G_0, G_1 is called *convex* if $G_0 \cap G_1$ is a convex subgraph of G. The expansion of G with respect to G_0, G_1 is called *peripheral* if $G_0 \subseteq G_1$ or $G_1 \subseteq G_0$.

3. Trivial and several sporadic examples

We have already mentioned that Q_3 is the only cubic median graph. Mulder [16,17] characterized median graphs as the graphs that can be obtained from K_1 by a sequence

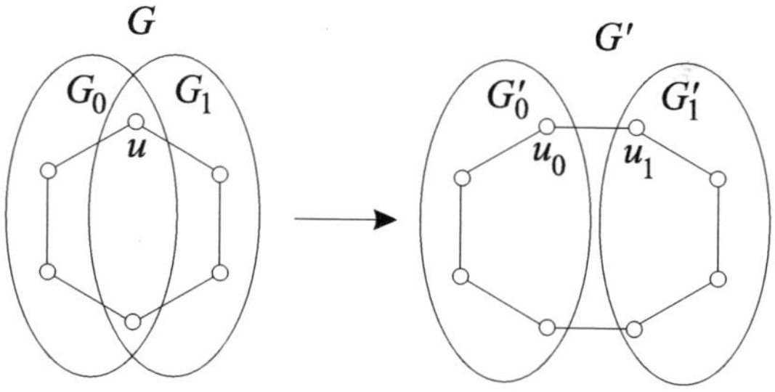

Figure 1. C_8 as an expansion of C_6

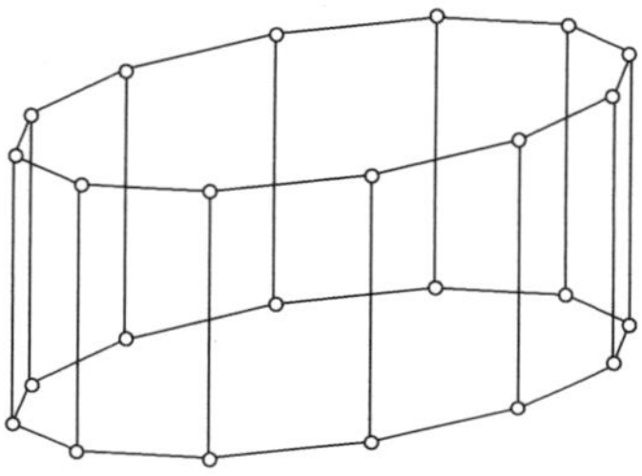

Figure 2. $K_2 \square C_{12}$

of convex expansions. In addition, the sequence of convex expansions in the characterization can be replaced by the so-called peripheral convex expansions [19].

On the other hand, Chepoi [4] proved that partial cubes are precisely the graphs obtainable from K_1 by a sequence of expansions. However, not every partial cube can be obtained by a peripheral expansion procedure. In [3] *tree-like partial cubes* were introduced as those partial cubes that can be obtained by a sequence of peripheral expansions from K_1. Do we get some more regular partial cubes among tree-like partial cubes? No:

Theorem 3.1 ([3]). *Regular tree-like partial cubes are hypercubes.*

Since Cartesian products of partial cubes are again partial cubes, the *even prisms*

$$K_2 \square C_{2n},\ n \geq 2,$$

form an infinite family of cubic partial cubes. They are called *trivial cubic partial cubes*. Note that the first graph of this series is Q_3. See Fig. 2 for $K_2 \square C_{12}$.

By an extensive computation, Bonnington, Klavžar, and Lipovec [1] determined that up to 30 vertices, there exist only 3 nontrivial examples of cubic partial cubes:

- the *Desargues graph*,
- a graph B_1 on 24 vertices (alias the permutahedron Π_3 [21, p.16]), and
- a sporadic example B'_1 on 30 vertices.

The three nontrivial examples of cubic partial cubes up to 30 vertices are shown in Fig. 3, the Desargues graph being the top one.

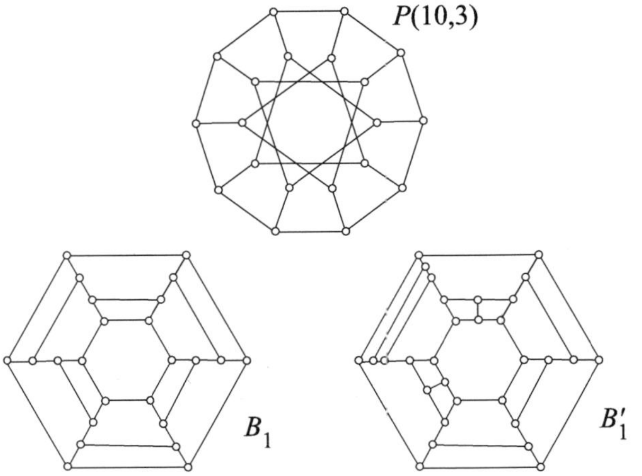

Figure 3. Cubic partial cubes on 20, 24 and 30 vertices

Knowing that the Desargues graph is a partial cube and that it can be described as the generalized Petersen graph $P(10, 3)$, it is natural to ask whether some additional generalized Petersen graphs are partial cubes.

Theorem 3.2 ([14]). *Prisms* $P(2n, 1) = K_2 \square C_{2n}$, $n \geq 2$, *and the Desargues graph* $P(10, 3)$ *are the only partial cubes among the generalized Petersen graphs* $P(n, k)$.

In [2] the following concept has been introduced in order to find more cubic partial cubes. Let G be a plane graph. (In fact, it could be embedded into an arbitrary closed surface, but until now, in this way no new cubic partial cubes were found.) Then replace each vertex $v \in V(G)$ by a cycle Q_v of length $2\deg_G(v)$, and replace every edge uv of G by two edges joining Q_u and Q_v in such a way that a cubic plane graph is obtained in which all cycles Q_v are facial and all edges of G give rise to 4-faces in that map. The resulting map $CI(G)$ is called the *cubic inflation* of G. Note that $CI(K_4)$ is just the permutahedron Π_3, a graph known to be a partial cube. Using the cubic inflation five new examples of cubic partial cubes were detected. The graphs $G_1, \ldots, G_5$ whose cubic inflations are partial cubes are shown in Fig. 4.

Deza, Dutour-Sikiric, and Shpectorov [5] classified plane cubic partial cubes that have only 4- and 6-gonal faces: there are precisely five such graphs:

- Q_3,
- $K_2 \square C_6$,
- the so-called truncated octahedron,
- the so-called chamfered cube, and
- the so-called twisted chamfered cube.

The investigations described in this section are briefly summarized in Table 1, where n stands for the number of vertices of the given graph.

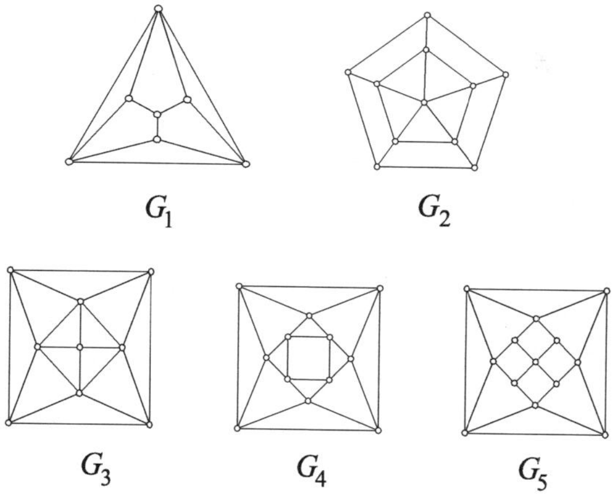

Figure 4. Graphs whose cubic inflations are partial cubes

Table 1. Several nontrivial cubic partial cubes.

n	graph		n	graph	
20	P(10,3)	[14,7]	48	$CI(Q_3)$	[6,7]
24	B_1	[1,7]	48	B_2	[1]
30	B_1'	[1]	80	$CI(G_2)$	[2]
32	chamfered cube	[5,7]	80	$CI(G_3)$	[2]
32	twisted chamfered cube	[5,7]	96	$CI(G_4)$	[2]
36	B_1''	[1]	96	$CI(G_5)$	[2]
42	B_1'''	[1]	120	CI(Icosahedron)	[6,7]
48	$CI(G_1)$	[2]			

4. Tribes of cubic partial cubes

In this section we briefly present a technique developed in [15] that led to many new examples of cubic partial cubes.

The expansion of a cubic graph G with respect to G_0, G_1 is a *cubic expansion* if $G_0 \cap G_1$ is a disjoint union of edges, where every vertex of $G_0 \cap G_1$ has one neighbor in $G_0 \backslash G_1$ and one neighbor in $G_1 \backslash G_0$. The reverse operation is called a *cubic contraction*. Now, if G is a cubic partial cube then every graph obtained from G by a cubic expansion or a cubic contraction is again a cubic partial cube. Therefore, we call a minimal nonempty collection of such graphs, closed under cubic expansions and cubic contractions, a *tribe* of cubic partial cubes. The tribe containing a particular graph G is called a *G-tribe*.

Considering the tribe of the trivial cubic partial cubes (that is, of the prisms) gives no new examples.

Proposition 4.1 ([15]). *The even prisms $K_2 \square C_{2n}$ form a tribe.*

The next tribe to be considered is the one of the smallest nontrivial cubic partial cube—the Desargues graph. Unfortunately, also in this case no new examples were found as the next result asserts.

Proposition 4.2 ([15]). *The generalized Petersen graph* $P(10, 3)$ *admits no cubic expansions and no cubic contractions. Consequently, the* $P(10, 3)$*-tribe consists only of* $P(10, 3)$ *itself.*

However, with the next graph, the graph B_1 from Table 1 (the left graph from Fig. 3), the situation dramatically changes. Let us rename B_1 to B_{24} in order to denote the size of this graph. Then the B_{24}-tribe consists of 17 graphs and is presented in Fig. 5. In the figure, an edge between two graphs means that one is obtained from the other by means of a cubic expansion or a cubic contraction.

The 9 squared graphs from Fig. 5 are new cubic partial cubes. The remaining graphs are included in Table 1; the graphs $B_{24}, B_{30}, B_{36}, B_{42}, B_{48}$ are the graphs $B_1, B_1', B_1'', B_1''', B_2$, respectively; B_{32} is the chamfered cube; B_{48} is the truncated cuboctahedron $CI(Q_3)$; B_{96}' is $CI(G_4)$. In Fig. 6 the graphs B_{36}' and B_{40} are shown.

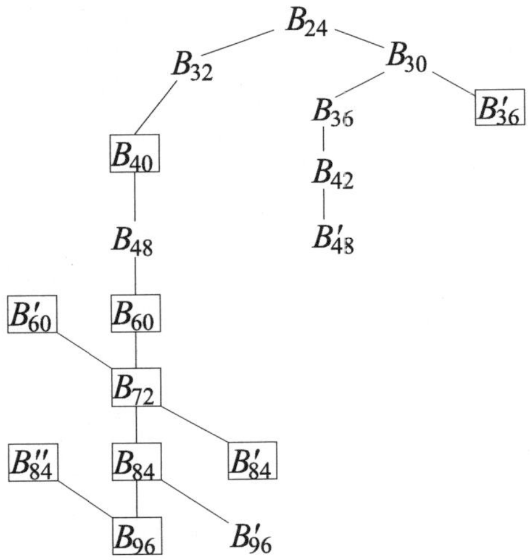

Figure 5. The B_{24}-tribe

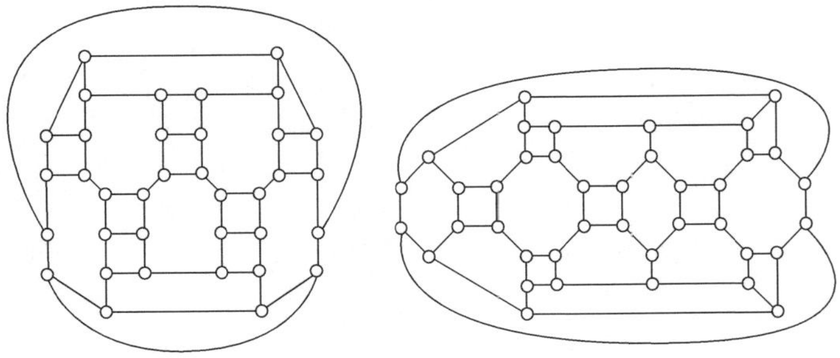

Figure 6. Two graphs from the B_{24}-tribe

The tribe of the twisted chamfered cube consists of two graphs, the second one being a new example on 40 vertices. Similarly, the $CI(G_1)$-tribe brings another new example of 60 vertices. The $CI(G_2)$-tribe consists of three graphs, and we get two new examples on 70 and 60 vertices that are obtained from $CI(G_2)$ by a cubic contraction. The $CI(G_3)$-tribe and the $CI(G_5)$-tribe both consist of a single graph, while the CI(Icosahedron)-tribe brings us 8 new cubic partial cube, the largest being on 240 vertices. For all these new examples see [15].

5. Cubic partial cubes from simplicial arrangements

Given a finite set of hyperplanes in $\mathbb{R}^d$, a partial cube G can be constructed as follows. Its vertices are the d-dimensional cells of the arrangement, two vertices being adjacent whenever the corresponding cells meet along a $(d-1)$-dimensional face of the arrangement. In this way, however, non-regular partial cubes are constructed. Instead, Eppstein [9] considered simplicial arrangements of lines in the real projective plane—finite sets of lines such that each of the cells is a triangle.

Theorem 5.1 ([9]). *To any simplicial line arrangement A in the projective plane there exists a cubic partial cube with twice as many vertices as A has triangles.*

Three infinite families of simplicial line arrangements are known [12]: the near-pencils, and families denoted by $R(2k)$ and $R(4k+1)$. Using Theorem 5.1 the near-pencils lead to the even prisms, while the families $R(2k)$ and $R(4k+1)$ give two new infinite families of cubic partial cubes.

Moreover, there are 91 additional sporadic examples of simplicial line arrangements [12], so they give numerous additional examples of cubic partial cubes.

A pseudoline arrangement is a collection of curves in the plane topologically equivalent to lines, such that any two curves cross in a single point. Such an arrangement is simplicial provided that each cell of it is bounded by sides belonging to three curves. Then Eppstein further proved:

Theorem 5.2 ([9]). *To any pseudoline arrangement A a partial cube can be associated. Moreover, if the arrangement is simplicial, the corresponding partial cube is cubic.*

Grünbaum mentioned in [12] that there are seven infinite families of simplicial arrangement of pseudolines, so we conclude that, after all, there is an abundance of cubic partial cubes. However, all these graphs are relatively large which could be an explanation why it took quite a while before they were detected.

6. Prospects

There seems to be a lot of work ahead. It would be nice to have a list (as complete as possible) of cubic partial cubes up to some number of vertices (say 100). For this sake, the concept of tribes seems very appropriate. One would need to compute all the tribes of the known cubic partial cubes, that is, those arising from the simplicial line arrangements of lines and pseudolines. In the latter case there seems to be technical problem. As far as we could found, there is no explicit list of the infinite families of pseudoline arrangements available.

As we already mentioned, the computation from [1] displays $P(10, 3)$ as **the** smallest nontrivial cubic partial cube. It seems rather surprising that among all discovered cubic partial cubes only the Desargues graph $P(10, 3)$ is not planar. Therefore we conclude with the following problem.

Problem 6.1. Is $P(10, 3)$ a unique nonplanar cubic partial cube?

The previous experience with cubic partial cubes indicate that non-planar cubic partial cubes are likely to be hiding somewhere out there. However, it seems that new techniques will have to be developed in order to find them.

References

[1] P. Bonnington, S. Klavžar and A. Lipovec, On cubic and edge-critical isometric subgraphs of hypercubes, *Australas. J. Combin.*, **28** (2003) 217–224.

[2] B. Brešar, S. Klavžar, A. Lipovec and B. Mohar, Cubic inflation, mirror graphs, regular maps, and partial cubes, *European J. Combin.*, **25** (2004) 55–64.

[3] B. Brešar, W. Imrich and S. Klavžar, Tree-like isometric subgraphs of hypercubes, *Discuss. Math. Graph Theory*, **23** (2003) 227–240.

[4] V. D. Chepoi, d-Convexity and isometric subgraphs of Hamming graphs, *Cybernetics*, **1** (1988) 6–9.

[5] M. Deza, M. Dutour-Sikiric and S. Shpectorov, Graphs 4_n that are isometrically embeddable in hypercubes, *Southeast Asian Bull. Math.*, **29** (2005) 469–484.

[6] M. Deza and V. P. Grishukhin, A zoo of l_1-embeddable polytopal graphs, *Bull. Inst. Math. Acad. Sinica*, **25** (1997) 181–231.

[7] M. Deza, V. Grishukhin and M. Shtogrin, Scale-Isometric Polytopal Graphs in Hypercubes and Cubic Lattices, Polytopes in Hypercubes and $\mathbb{Z}_n$, Imperial College Press, London (2004).

[8] A. A. Dobrynin, I. Gutman, S. Klavžar and P. Žigert, Wiener index of hexagonal systems, *Acta Appl. Math.*, **72** (2002) 247–294.

[9] D. Eppstein, Cubic partial cubes from simplicial arrangements, *Electron. J. Combin.*, **13** (2006) #R79.

[10] K. Fukuda and K. Handa, Antipodal graphs and oriented matroids, *Discrete Math.*, **111** (1993) 245–256.

[11] R. L. Graham and H. Pollak, On the addressing problem for loop switching, *Bell System Tech. J.*, **50** (1971) 2495–2519.

[12] B. Grünbaum, Arrangements and Spreads, American Mathematical Society, Providence, R.I. (1972).

[13] S. Klavžar, I. Gutman and B. Mohar, Labeling of benzenoid systems which reflects the vertex–distance relations, *J. Chem. Inf. Comput. Sci.*, **35** (1995) 590–593.

[14] S. Klavžar and A. Lipovec, Partial cubes as subdivision graphs and as generalized Petersen graphs, *Discrete Math.*, **263** (2003) 157–165.

[15] S. Klavžar and S. Shpectorov, Tribes of cubic partial cubes, *Discrete Math. Theor. Comput. Sci.*, **9** (2007) 273–292.

[16] H. M. Mulder, The structure of median graphs, *Discrete Math.*, **24** (1978) 197–204.

[17] H. M. Mulder, The Interval Function of a Graph, Math. Centre Tracts 132, Mathemtatisch Centrum, Amsterdam (1980).

[18] H. M. Mulder, n-Cubes and median graphs, *J. Graph Theory*, **4** (1980) 107–110.

[19] H M. Mulder, The expansion procedure for graphs, in: Contemporary Methods in Graph Theory, (ed.) R. Bodendiek, B.I.-Wissenschaftsverlag, Manhaim/Wien/Zürich (1990) 459–477.

[20] S. Ovchinnikov, Hyperplane arrangements in preference modeling, *J. Math. Psych.*, **49** (2005) 481–488.

[21] G. M. Ziegler, Lectures on Polytopes, Springer, New York (1998).

Joint Proc. Int. Wrk.shps. Conv. in Discr. Struct. and Metric & Conv. Graph Th. No. 5, 2008, pp. 97–104.

Convexity of Minimal Dominating Functions and Minimal Edge Dominating Functions

K. Reji Kumar

Department of Mathematics, N.S.S College, Pandalam 689 501, India
e-mail: rkkmaths@yahoo.co.in

Abstract. In this paper we study fractional edge domination of the graph $G = (V, E)$ as an analogue to fractional domination. Convexity of minimal edge dominating functions are investigated and certain classes of graphs having universal minimal edge dominating functions are given.

Keywords. Minimal dominating function, Minimal edge dominating function, Universal minimal edge dominating function.

2000 Subject Classification: Primary 05C 69, Secondary 05C 35

1. Introduction

The notation $G = (V, E)$ represents a finite, undirected graph without loops nor multiple edges. Readers may depend Harary [6] or Douglas. B. West [11] for basic reference. In this section we shall discuss the background of the concepts studied.

Even though Farber [4] introduced indirectly the concept of fractional domination in graphs, the subject was studied in depth by Hedetniemi *et al.* [9]. Almost all details of latest developments in the area of domination in graphs and related topics are collected by T. W Haynes *et al.* in [7] and [8]. If g is a function mapping the vertex set V into some set of real numbers and $S \subseteq V$, then $g(S) = \sum_{v \in S} g(v)$ and the *weight* $|g| = g(V)$. A real valued function $f : V \to [0, 1]$ is a *dominating function* (DF) of $G = (V, E)$ if for every $v \in V$,

$$\sum_{x \in N[v]} f(x) \geq 1.$$

Here $N[v]$ is the closed neighbourhood of the vertex v. Thus if S is a dominating set of vertices of G, then χ_S is a dominating function. A DF f is called *minimal dominating function* (MDF) [2] if there is no function $g : V \to [0, 1]$ such that $g < f$ and g is a DF. The set of all minimal dominating functions is denoted by $\mathfrak{D}$. The *fractional domination number*, denoted $\gamma_f(G)$ and the *upper fractional domination number*, denoted $\Gamma_f(G)$, are defined as follows.

$$\gamma_f(G) = \min\{|g| : g \text{ is a minimal dominating function of } G\}$$
$$\Gamma_f(G) = \max\{|g| : g \text{ is a minimal dominating function of } G\}$$

For a DF f of G, the *boundary* set B_f and the *positive set* P_f are defined by

$$B_f = \{v \in V / f(N[v]) = 1\}$$

and

$$P_f = \{v \in V / f(v) > 0\}.$$

Let A and B be subsets of V. We say that A dominates B and write $A \to B$ if every vertex in $B - A$ is adjacent to some vertex in A. Following theorem gives a simple necessary and sufficient condition for a DF to be an MDF.

Theorem 1.1 ([2]). *A dominating function f is minimal if and only if $B_f \to P_f$.*

Majumdar raised the following question [2], which is considered as the motivation behind the definition of the convex combination of two MDFs.

Suppose that, f and g are two MDFs of G and $f(V) < x < g(V)$. Does there exists an MDF of G with aggregate x?

For the real valued functions f and g defined on V and $\lambda \in (0, 1)$, the convex combination h_λ of f and g is defined by $h_\lambda(v) = \lambda f(v) + (1 - \lambda) g(v)$ for each $v \in V$. Convex combinations of MDFs are DFs, but not minimal always. This motivated to study the convexity properties of MDFs.

Theorem 1.2 ([2]). *Any convex combination of two MDFs f and g is minimal if and only if $B_f \cap B_g \to P_f \cup P_g$.*

If h_λ is minimal for some $\lambda \in (0, 1)$, then it is minimal for all $\lambda \in (0, 1)$. We write $f R g$, if h_λ is minimal. This relation is always reflexive and symmetric but not transitive, which implies the set of all MDFs is not convex always. If f is an MDF of the graph G such that $f R g$ for any MDF g of G, then f is called a *universal* MDF (UMDF) of G. The set of all MDFs is convex if and only if all MDFs are universal MDFs.

The *line graph* $L(G) = (V', E')$ of a given graph G is a graph such that $V' = E(G) = \{e_1, e_2, \ldots, e_n\}$ and $E' = \{(e_i, e_j) : e_i$ and e_j are incident, where $i \neq j\}$. The following characterization of line graphs is used in the sequel.

Theorem 1.3 ([6]).

(1) *G is a line graph if and only if lines of G can be partitioned into complete subgraphs in such a way that no point lies in more than two of the subgraphs.*
(2) *Let G and G' be connected graphs with isomorphic line graphs. Then G and G' are isomorphic unless one is K_3 and the other is $K_{1,3}$.*

Suppose G is a line graph. Then by Theorem 1.3 (1), its edges can be partitioned into complete subgraphs in such a way that no vertex belongs to more than two of them and by Theorem 1.3 (2), the partition is unique unless the line graph is K_3. We call a vertex *intersecting* if it belongs to exactly two of these complete subgraphs. Let u be a vertex of the graph H. A *complete addition* [2] to H at u is the identification of u and a vertex of some complete graph having at least two vertices.

Theorem 1.4 ([2]). *Let H be any graph. From H, make G by one or more complete additions to H at u for each vertex u of H. Then each MDF of G is universal.*

A vertex $v \in V$ is said to *absorb* the vertex $u \neq v$, and u is said to *be absorbed* by v, if $N[u] \subset N[v]$, where $\subset$ denotes strict inclusion. The vertex v is also called an *absorbing vertex* and u is called the *absorbed* vertex. Let

$$A = \{v \in V(G) : v \text{ is an absorbing vertex }\}$$

We define a vertex w of G to be $f-sharp$, where f is an MDF of G, if $B_f \cap N[w] \subseteq A$, and w is said to be *sharp* if it is f-sharp for some MDF f of G. Absorbing vertices and sharp vertices play an important role in the characterization of UMDFs of a given graph.

By replacing the closed neighbourhood of the vertex v in the definition of MDF, by its open neighbourhood $N(v)$, we get *minimal total dominating function* (MTDF) of G. *Low* vertices – used in the theory of total dominating functions (TDFs) – are analogues to sharp vertices used in the theory of DFs. Its definition is very similar to that of sharp vertices.

Theorem 1.5 ([1]). *The MDF g of G is a universal MDF if and only if*

(i) $V - A \subseteq B_g$ *and*
(ii) $g(w) = 0$ *for each sharp vertex w of G.*

Theorem 1.6 ([3,5]). *For every graph with n vertices, maximum degree $\Delta(G)$ and minimum degree $\delta(G)$, $\frac{n}{(\Delta(G)+1)} \leq \gamma_f(G) \leq \frac{n}{(\delta(G)+1)}$.*

Corollary 1.7 ([3,5]). *If G is an r regular graph with n vertices,*

$$\gamma_f(G) = \frac{n}{r+1}.$$

Theorem 1.8 ([10]). *For the graphs K_m where $m \geq 0$, $\gamma_f = \Gamma_f = 1$.*

2. Minimal dominating functions

The first result in this section is the characterization of the vertex-transitive graphs whose set of MDFs is a convex set. This is obtained by modifying the Theorem 1.5.

Lemma 2.1. *All MDFs of G are universal MDFs if and only if*

(i) $V - A \subseteq \cap_{f \in \mathfrak{D}(G)} B_f$ *and*
(ii) *G has no sharp vertices.*

Proof. Since the Theorem 1.5 is true for all $f \in \mathfrak{D}(G)$, rest of the proof is obvious and hence omitted. □

A *vertex-transitive* graph is a graph G such that, given any two vertices v_1 and v_2 of G, there is some automorphism $f : V \to V$ such that $f(v_1) = v_2$. The graphs K_n where $n \geq 2$ and C_n where $n \geq 3$ are suitable examples. It is very interesting to note that, in these graphs, all vertices have equal importance. In such graphs, if a vertex v is an absorbing vertex, then all vertices are absorbing. Similarly if G contains a low vertex, then all vertices are low vertices.

Theorem 2.2. *The only vertex-transitive graphs with all MDFs universal are the complete graphs.*

Proof. If G is a vertex-transitive graph, then all vertices of G are equally important. Hence G has no absorbing vertices, sharp vertices and low vertices. Otherwise all vertices are absorbing, sharp and low vertices.

Next suppose G is vertex transitive but not complete. For any vertex x there exist non-adjacent vertices v and w in $N(x)$. The characteristic function of a maximal independent set containing v and w is an MDF and x is not in its boundary. Hence the intersection of the boundaries of all MDFs is empty. If G is a complete graph, since $N[v] = V$ for all $v \in V(G)$, the boundary of all MDFs of G is V. Hence the result follows from Lemma 2.1 (i). □

Example 2.3. The set of all MDFs of C_n where $n \geq 3$ is not a convex set. For C_n where n is even, having $V(C_n) = \{v_1, v_2, \ldots, v_n\}$ define the function,

$$f(v_i) = \begin{cases} 1 & \text{if } i \text{ is odd} \\ 0 & \text{if } i \text{ is even} \end{cases}$$

For C_n where n is odd, define

$$g(v_i) = \begin{cases} 1 & \text{if } i \text{ is odd and } i \neq n \\ 0 & \text{if } i \text{ is even or } i = n \end{cases}$$

The functions f and g are MDFs but $B_f \neq V$ and $B_g \neq V$. This implies, the set of all MDFs of C_n is not a convex set.

3. Minimal edge dominating functions

Even though fractional version of domination, independence, irredundance etc. are well studied for vertex sets, their counterpart for edge sets are totally ignored. Motivated by this we define fractional edge domination, which is analogous to dominating functions. At the end of this section, we shall establish a direct relation between these functions and the corresponding vertex functions through the line graph of a given graph.

In a graph $G = (V, E)$, let $e = (v_1, v_2) \in E$, the edges $x_1, x_2, \ldots x_n$ and the edges $y_1, y_2, \ldots y_m$ are incident at v_1 and v_2 respectively, then the closed neighbourhood of e is, $N[e] = \{e, x_1, x_2, \ldots x_n, y_1, y_2, \ldots y_m\}$. For the function $f : E \to \mathbb{R}$ and $S \subseteq E$, $f(S) = \sum_{e \in S} f(e)$.

Definition 3.1. *The real valued function $f : E \to [0, 1]$ is called an* edge dominating function *(EDF) if $\sum_{x \in N[e]} f(x) \geq 1$ for all $e \in E$, where $N[e]$ is the closed neighbourhood of e. An EDF f is called a* minimal edge dominating function *(MEDF) if there is no function $g : E \to [0, 1]$ such that $g < f$ and g is an EDF. The set $\mathfrak{D}'(G) = \{f : f$ is an MEDF of the graph $G\}$. The* fractional edge domination number, *denoted by $\gamma'_f(G)$ and the* upper fractional edge domination number, *denoted by $\Gamma'_f(G)$, are defined as follows.*

$$\gamma'_f(G) = \min\{|g| : g \text{ is a minimal edge dominating function of } G\}$$

$$\Gamma'_f(G) = \max\{|g| : g \text{ is a minimal edge dominating function of } G\}$$

The relation

$$\gamma_f(L(G)) = \gamma'_f(G) \leq \Gamma'_f(G) = \Gamma_f(L(G))$$

is used to find these parameters for certain classes of graphs. Every minimal edge dominating function of a graph G is a minimal dominating function of the line graph $L(G)$. Hence the counterparts of all major results proved in the theory of dominating functions are true in the present situation. The most important results which are relevant in the sequel are given below.

The *boundary* set B'_f and the *positive set* P'_f of an EDF are defined by

$$B'_f = \{e \in E / f(N[e]) = 1\}$$

and

$$P'_f = \{e \in E / f(e) > 0\}.$$

For two subsets A and B of E, we say that A dominates B and write $A \rightsquigarrow B$ if every edge in $B - A$ is adjacent to some edge in A.

Theorem 3.2. *An EDF f of a graph G is an MEDF if and only if $B'_f \rightsquigarrow P'_f$.*

Remark 3.3. An edge function of a particular type, of a graph G can give a vertex function of the same type of the line graph $L(G)$ and vice versa. Hence all properties of edge dominating functions can be studied by analyzing the corresponding vertex functions of $L(G)$.

Theorem 3.4. *For the complete graph K_n where $n \geq 2$ and $n \neq 8$, $\gamma'_f(K_n) = \frac{n(n-1)}{2(2n-3)}$.*

Proof. Line graph of K_n where $n \geq 2$ and $n \neq 8$ has been characterized and proved that it contains ${}_nC_2$ vertices and is regular of degree $2(n-2)$. Hence by Corollary 1.7, the result follows. □

Theorem 3.5. *Unless $m = n = 4$, $\gamma'_f(K_{(m,n)}) = \frac{mn}{(m+n-1)}$.*

Proof. Line graph of $\gamma'_f(K_{(m,n)})$, except when $m = n = 4$ is regular of degree $m+n-2$ and contains mn points. Now by applying Corollary 1.7, we get the result. □

Least possible value of γ'_f is 1. Star graph is an example. Next we attempt to characterize the graphs having $\gamma'_f = 1$. Consider the class of all graphs ($\mathfrak{M}$) of the following type,

(i) $G = K_{(1,m)}$ where $m \geq 0$.
(ii) Any graph G formed by identifying the vertex of $K_{(1,m)}$ having maximum degree with the vertex of $K_{(1,n)}$ having degree one, where the integers $m, n \geq 0$.
(iii) Any graph G, obtained from any graph of type (ii) by identifying one or more pairs of pendant vertices, so that no multiple edges are formed.

Now we proceed to prove the characterization of graphs having $\gamma'_f = 1$.

Theorem 3.6. *For the graph $G = (V, E)$, $\gamma'_f = 1$ if and only if $G \in \mathfrak{M}$.*

Proof. We shall prove the theorem in two steps.

1. $\gamma'_f = 1$ if and only if $L(G)$ has a dominating vertex
2. The graphs having line graph, with a dominating vertex are precisely those described in (i), (ii) and (iii).

To prove (1), it is enough to show that $L(G)$ has an MDF f with $\gamma_f = 1$ if and only if $L(G)$ has a dominating vertex (from which 1 follows). If $\gamma_f = 1$ then $\sum_{x\in V} f(x) = 1 = \sum_{w\in N[u]} f(w)$ for every vertex u. Therefore every vertex of $L(G)$ is adjacent to every vertex of P_f, so every vertex of P_f is dominating. To prove the converse, if x is a dominating vertex, then set $f(x) = 1$ and $f(y) = 0$ for all $y \in (V - \{x\})$. The function f is a minimal dominating function and $\sum_{x\in V} f(x) = 1$.

To prove (2), note that $L(G)$ has a dominating vertex if and only if either there is a vertex x such that all edges are incident with x (case (i)), or there are adjacent vertices x and y such that every edge of G is incident with x or y. Then $N(x) - N(y)$ and $N(y) - N(x)$ must both consist of vertices of degree 1, and $N(x) \cap N(y)$ must consist of vertices of degree 2. These are the graphs described in (ii) and (iii), with (ii) being the case that $N(x) \cap N(y)$ is empty. □

Remark 3.7. Suppose we apply complete addition at every vertex of a line graph L, exactly once and get a new graph L'. This graph is also a line graph, if and only if $L = K_n$ where $n \geq 0$. If L is any graph other than K_n, then by Theorem 1.3 (1), we can find at least one vertex $u \in V(L)$ such that u is common to two complete graphs. A complete addition at u increases the number of complete graphs at u and again by Theorem 1.3 (1), L' is not a line graph. It is quite easy to see that L' is a line graph, if $L = K_n$.

Let $\mathfrak{A}$ denote the class of all line graphs $L = K_n$ where $n \geq 0$ and L' formed from $L = K_n$ as stated in Remark 3.7. The *convex combination* of two MEDFs and the *universal minimal edge dominating function* (UMEDF) are defined in exactly same way as that of vertex domination. In the following lines we see some classes of graphs having convex set of MEDFs.

Theorem 3.8. *Let G be a graph such that $L(G) \in \mathfrak{A}$. Then each MEDF of G is a universal MEDF.*

Proof. By the Theorem 1.4 and the Remark 3.7, each MDF of $L(G)$ is a universal MDF. Hence each MEDF of G is a universal MEDF. □

Theorem 3.9. *Let $L(G)$ be the line graph of G satisfying the conditions*

(i) *$L(G)$ has only intersecting vertices.*
(ii) *$L(G)$ has no sharp vertices.*
(iii) *If K_l, K_m and K_n are induced subgraphs of $L(G)$ having mutual intersections, then $l, m, n > 1$.*

Then g is a universal minimal edge dominating function of the graph G, if and only if $B'_g = E(G)$.

Proof. First we claim that, if the line graph $L(G)$ satisfies the given conditions, then the set of all absorbing vertices of $L(G)$, $A(\mathrm{L(G)}) = \phi$. To prove the claim, suppose that $A \neq \phi$ and $v \in V(L(G))$ is an absorbing vertex. Then by the definition of absorbing vertex, there exists a vertex $u \in V(L(G))$, and $N[u] \subset N[v]$. Since every vertex is an intersecting vertex, the vertex v is common to two complete graphs, say K_l and K_m. By the characterization of line graphs, $u \in V(K_l)$ or $u \in V(K_m)$. Both cases cannot be possible simultaneously. Since u is an intersecting vertex, there exists K_n such that, u is common to K_m and K_n. This is possible, if and only if $m = 1$, since $N[u] \subset N[v]$. Now by condition (ii), we get the contradiction. Hence the claim is proved.

Thus we get $V(L(G)) - A = V(L(G))$. Now by Theorem 1.5, an MDF h of $L(G)$ is a universal MDF if and only if $V(L(G)) = B_h$. This is equivalent to $B'_g = E(G)$, where g is the corresponding minimal edge dominating function. □

Example 3.10. Consider the cycles C_n where $n > 3$. Line graph of C_n is the same graph. This line graph satisfies all conditions given in Theorem 3.9. The function $g : E(C_n) \to [0, 1]$ such that $g(e) = \frac{1}{3}$ for all $e \in E$ is a universal minimal edge dominating function, since $B'_g = E(G)$.

Remark 3.11. It has been proved by Cockayne *et al.* [2] that the cycle C_3 and the complete graph K_n have universal MDFs. Since these graphs are line graphs, its associated graphs, respectively C_3 and the star graphs have universal MEDFs.

Theorem 3.12. *The graph G formed by identifying the vertex of $K_{(1,m)}$ having maximum degree with the vertex of $K_{(1,n)}$ having degree one, where the integers $m, n \geq 0$ has universal MEDF.*

Proof. Let $L(G)$ be the line graph of G such that $V(L(G)) = \{v, v_1, v_2, \ldots, v_{(m-1)}, u_1, u_2, \ldots, u_{(n-1)}\}$. The vertex v is common to the complete graphs K_m and K_n.

Let f be the characteristic function of $\{v\}$. Then $B_f = V$ and $P_f = \{v\}$. By Theorem 1.1, f is an MDF. Let g be any other MDF of $L(G)$. Then $g(v) < 1$. Since $g(N[v_i]) \geq 1$ for $i = 1, 2, \ldots, m$ and $g(N[u_j]) \geq 1$ for $j = 1, 2, \ldots, n$ and $B_g \to P_g$, $v_i, u_j \in B_g$ for all i and j. Hence $B_g = V - \{v\}$. If h is a convex combination of f and g, then $B_h = B_f \cap B_g = B_g = V - \{v\}$ and $P_h = P_f \cup P_g = P_g \cup \{v\}$. It is not difficult to show that $B_h \to P_h$ and by Theorem 1.2, h is an MDF. Hence f is a universal MDF of $L(G)$. □

The only class of graphs having $\Gamma_f = 1$ is K_n where $n \geq 1$. These graphs have the property, $\gamma_f = \Gamma_f = 1$. This discussion leads to the following theorem which is the characterization of graphs having the property, $\gamma'_f = \Gamma'_f = 1$.

Theorem 3.13. *Let G be a graph, then $\gamma'_f = \Gamma'_f = 1$ if and only if $G = K_{(1,m)}$ where $m \geq 0$ or $G = K_3$.*

Proof. First we shall show that, $\gamma_f(H) = \Gamma_f(H) = 1$ if and only if H is a complete graph. If H is not complete, then it has an independent set X of cardinality $k > 1$. The characteristic function of X is an MDF with aggregate $k > 1$. Hence $\Gamma_f(H) > 1$. Now the result follows from the fact that, $L(G)$ is complete if and only if G is a star or $G = K_3$. □

4. Problems for further study

I think that the following set of interesting problems related to the concept of minimal edge dominating functions can motivate the readers for further research.

(1) Is it possible to characterize the graphs having the set $\mathfrak{D}'(G)$, convex?
(2) Find new classes of graphs having UMEDFs.
(3) Find the characterization of graphs having UMEDFs.
(4) Develop suitable algorithms to find all MEDFs of a given graph.
(5) Determine γ'_f and Γ'_f for standard families of graphs.
(6) How the inclusion or deletion of vertices and edges to a graph G affects the convexity of the sets $\mathfrak{D}(G)$ and $\mathfrak{D}'(G)$.

Acknowledgement

I express my sincere gratitude to the anonymous referee for his valuable suggestions, that helped me to improve the quality of the paper. Also I am thankful to Dr. Manoj Changat for giving me an opportunity to present this paper in the International Workshop on Convexity of Discrete Structures, organized by The Department of Future studies, University of Kerala, Thiruvananthapuram.

References

[1] E. J. Cockayne, G. MacGillivray, and C. M. Mynhardt, Convexity of minimal dominating functions and universals of graphs, *Bull. Inst. Combin. Appl.*, **5** (1992) 37–48.
[2] E. J. Cockayne, G. Fricke, S. T. Hedetniemi and C. M. Mynhardt, Properties of minimal dominating functions of graphs, *ARS Combinatoria.*, **41** (1995) 107–115.
[3] G. S. Domke, S. T. Hedetniemi and R. C. Laskar, Fractional packing covering and irredundance in graphs, *Congr. Numer.*, **66** (1988) 227–238.
[4] M. Farber, Domination, independent domination and duality in strongly chordal graphs, *Discrete. Appl. Math.*, **7** (1986) 115–130.
[5] D. Grinsted and P. J. Slater, Fractional domination and fractional packing in graphs, *Congr. Numer.*, **71** (1990) 153–172.
[6] F. Harary, Graph Theory, Addison-Wesley, Reading, Mass (1972).
[7] T. W. Haynes, S. T. Hedetniemi and P. J. Slater, Fundamentals of Domination in Graphs, Marcel Dekker, Inc., New York (1998).
[8] T. W. Haynes, S. T. Hedetniemi and P. J. Slater, Domination in Graphs – Advanced Topics, Marcel Dekker, Inc., New York (1998).
[9] S. M. Hedetniemi, S. T. Hedetniemi and T. V. Wimer, Linear time resource allocation algorithms for trees, Technical report URI-014, Department of Mathematics, Clemson University (1987).
[10] K. Reji Kumar, Studies in Graph Theory-Dominating functions, Ph.D. thesis, Manonmaniam Sundaranar University, Tirunelveli, India (2004).
[11] D. B. West, Introduction to Graph Theory, Prentice Hall; 2 edition (2000).

Joint Proc. Int. Wrk.shps. Conv. in Discr. Struct. and Metric & Conv. Graph Th. No. 5, 2008, pp. 105–116.

Transit Functions on Posets and Lattices*

Antony Mathews[1] and Joseph Mathews[2]

[1]Department of Futures Studies, University of Kerala, Thiruvananthapuram 695 034, India
[2]Department of Mathematics, S.B. College, Changanassery 686 101, India
e-mail: sonykandans@yahoo.co.in; jose_chingam@yahoo.co.in

Abstract. Like the study of the transit functions on graphs, transit functions on posets and lattices are studied. A transit function R defined on a non empty set V is a betweenness if it satisfies the two betweenness axioms: $(b1)$ if $x \in R(u, v)$ and $x \neq v$ then $v \notin R(u, x)$ for u, v in V. $(b2)$ if $x \in R(u, v)$ then $R(u, x) \subseteq R(u, v)$ for u, v in V. R is said to satisfy the monotone axiom if $R(x, y) \subseteq R(u, v)$ for every $x, y \in R(u, v)$. The betweenness and monotone properties of the natural transit functions on a poset, a semi lattice and a lattice are studied. Lattices for which the lattice transit function satisfies the betweenness axioms are characterized by forbidden sublattices.

Keywords. Transit function, poset, lattice, cover, betweenness, monotonicity, covering graph, underlying graph.

Subject Classification: 06A06, 06B99, 03G10, 05C62, 05C99, 52A01

1. Introduction

The notions of intervals, betweenness and convexity are inherent in the usual interval in the real line and these natural concepts are axiomatized by Calder [7], known as interval function to study the convexities associated with this function. But the axioms of the interval function are so loose that it may not possess the natural betweenness properties like that of the real line. In the case of discrete structures, these functions give a rule to move around in the structure, for example the interval function in a connected graph is used by many authors as a fundamental tool in metric graph theory, see the important references, [1–6, 12–14, 17, 21]. Because of the very general properties of such functions, Mulder termed this function as 'transit function', see [20].

To be precise, a discrete structure (V, σ) consists of a nonempty set V together with a structure σ. Some pertinent structures are graphs, posets, lattices etc. A transit function is a function satisfying three simple axioms called transit axioms on (V, σ). The general idea is to consider a discrete structure (V, σ) and a transit function R defined on V and study the interrelations between σ and R. For example, transit functions defined in terms of paths in the graph $G = (V, E)$ are called path transit functions on G. Well studied path transit functions on graphs are the geodesic [20,21], induced path [9,10,19] and all paths [8] transit functions. For a survey on path transit functions, see

*Research supported by DST under the grant DST/INT/SLOV/P-03/05. The research of Antony Mathews is also supported by UGC under the FIP scheme.

[11]. In this paper we make an attempt to study transit functions on posets, lattices and semilattices.

Note that as for transit functions, a betweenness can just be defined on a set V by ignoring the structure σ of the set V. Thus one can develop an abstract theory of betweenness as in [23]. The geodesic transit function is a betweenness and has been studied by Mulder [21] in detail. The notion of betweenness in the above sense was first introduced in [19] where the graphs for which the induced path transit function satisfying betweenness axioms have been characterized by forbidden induced subgraphs. Any transit function on (V, σ) defines a natural convexity on V. See Van de vel [25] for a thorough discussion on various aspects of convexities (especially on interval convexities). The study of transit functions on posets is a new beginning, nevertheless there are many studies on order convexity on posets, lattices and similar structures. See for example [25 – 28] for detailed discussion of order convexities. The concept of order convexity is attributed to Birkhoff [15] and convexity in general (semi) lattices has been studied by Varlet [24] and Jamison [18].

If $P = (V, \leq)$ is a poset, the poset transit function R_P is the prime example of a transit function defined on P. Other examples are the meet (join) semilattice transit function R_S defined on a meet (join) semilattice $S = (V, \leq)$, the lattice transit function R_L defined on a lattice $L = (V, \leq)$.

In this paper we study the betweenness and monotone properties of the order transit function R_P, the meet (join) semilattice transit function R_S and the lattice transit function R_L and characterize, using forbidden sublattices, the lattices for which the lattice transit function R_L satisfies the betweenness axioms. Also studied is the relation of the underlying graphs of the poset transit function R_P, the semilattice transit functions R_S and the lattice transit function R_L respectively with the corresponding covering graphs of P, S and L. In this way, we make an attempt to connect graphs and posets using transit functions.

2. Preliminaries

A *transit function* on a non empty set X is a function $R_X : X \times X \to 2^X$ satisfying the three transit axioms:

$$(t1) \quad u \in R(u, v) \text{ for any } u \text{ and } v \in X.$$

$$(t2) \quad R(u, v) = R(v, u) \text{ for all } u \text{ and } v \in X.$$

$$(t3) \quad R(u, u) = \{u\} \text{ for all } u \in X.$$

A transit function R on X is a *betweenness* if it satisfies the betweenness axioms:

$$(b1) \quad x \in R(u, v);\, x \neq v \implies v \notin R(u, x) \text{ for } u, v \text{ in } X.$$

$$(b2) \quad x \in R(u, v) \implies R(u, x) \subseteq R(u, v) \text{ for } u, v \text{ in } X.$$

The axioms $(b1)$ and $(b2)$ seem to be very natural axioms of betweenness. We may also consider extra axioms and obtain special types of betweenness. One such axiom is

$$(b3) \quad \text{If } a, b \in X \text{ and } c, d \in R(a, b) \text{ then } c \in R(a, d) \implies d \in R(c, b).$$

A transit function R according to Van de vel and Nebesky is *geometric* if it satisfies $(b2)$ and $(b3)$. See [22,25]. One can show that, for a transit function, $(b3)$ implies $(b1)$, that is geometric implies betweenness but not conversely.

A transit function R is said to satisfy the *monotone axiom* if $R(x, y) \subseteq R(u, v)$ for every $x, y \in R(u, v)$. The *underlying graph* G_R of a transit function R on a set X is the graph with vertex set V where distinct u and v in V are joined by an edge if and only if $|R(u, v)| = 2$.

A *partially ordered set or poset* $P = (V, \leq)$ consists of a nonempty set V and a reflexive, anti-symmetric, transitive relation $\leq$ on V. If $u \leq v$ in a poset, we say u and v are *comparable*. If $u \leq v$ but $u \neq v$, then we write $u < v$. If u and v are in V, then v *covers* u in P if $u < v$ and there is no w in V with $u < w < v$. Using the covering relation we can obtain a graphical representation of P, called the *Hasse diagram* of P, as follows: draw a small circle to represent each element of P, by placing v higher than u whenever $u < v$; draw a straight line joining u and v whenever v covers u, otherwise join them by dotted lines. The figure $1(a)$, (b) are Hasse diagrams of posets. We can consider that the Hasse diagram of a poset is oriented either downwards or upwards. If the partial order '$\leq$' is a total order on V, then the poset P is called a *totally ordered set* or *chain*.

The *greatest lower bound or meet* of two elements u and v in P is an element x such that $x \leq u$ and $x \leq v$ and for any y in V with $y \leq u$ and $y \leq v$, we have $y \leq x$. If such a greatest lower bound of u and v exists, then it is unique, by the antisymmetry of $\leq$, and we denote it by $u \wedge v$. Similarly the *least upper bound or join* of u and v is an element x in V such that $u \leq x$ and $v \leq x$ and for any y in V with $u \leq y$ and $v \leq y$, we have $x \leq y$. If such a least upper bound exists, then it is unique, and we denote it by $u \vee v$. A *lattice* is a poset P in which any two elements have a meet and a join. A *meet (join) semilattice* is a poset P in which any two elements have a meet (join). In fact the posets represented by the figures (1) and (2) are lattices. A nonempty subset L_1 of a lattice L is called a *sublattice* of L if the meet and join of any two elements of L_1 is in L_1. The *covering graph* $G(P)$ of a poset $P = (V, \leq)$ is the graph with V as vertex set, in which two vertices are joined by an edge whenever one of the two vertices covers the other in the poset. In other words the covering graph of a poset is the unoriented Hasse diagram of the poset. Let $P = (V, \leq)$ and $Q = (W, \leq)$ be two posets. A one-one onto map $f : P \to Q$ is an *isomorphism* if $u \leq v \Leftrightarrow f(u) \leq f(v), \forall u, v \in V$ and in that case we write $P \cong Q$. If P and Q are lattices, then $P \cong Q$ if f satisfies the additional conditions:

(1) $f(u \wedge v) = f(u) \wedge f(v), \forall u, v \in V.$
(2) $f(u \vee v) = f(u) \vee f(v), \forall u, v \in V.$

If P and Q are meet or join semilattices, then $P \cong Q$ if f satisfies (1) or (2) respectively. L is *free* from K means that L has no sublattice isomorphic to K. For references on posets and related concepts see [16]. In this paper we denote a poset by P, a semilattice by S and a lattice by L and their underlying set by V and the partial ordering by '$\leq$'.

3. Transit functions on posets and semilattices, their betweenness

Definition 1. *For a poset P, the poset transit function $R_P : V \times V \to 2^V$ is defined as*

$$R_P(x, y) = \{x, y\} \text{ if } x \text{ and } y \text{ are incomparable}$$

$$R_P(x, y) = \{z|x \leq z \leq y\} \text{ if } x \leq y.$$

For a meet semilattice S, the meet semilattice transit function $R_S : V \times V \to 2^V$ is defined as $R_S(u, v) = \{z|u \wedge v \leq z \leq u$ or $u \wedge v \leq z \leq v\}$. In a similar way the join semi lattice transit function is defined as $R_S(u, v) = \{z|u \leq z \leq u \vee v$ or $v \leq z \leq u \vee v\}$.

We have the following proposition.

Proposition 1.

(i) *R_P is a betweenness on P.*
(ii) *R_S is a betweenness on S.*

Proof.

(i) Consider $R_P(u, v)$ where $u, v \in V$. To show that R satisfies $(b1)$, we consider two cases.

Case 1. Suppose u and v are incomparable. Then by definition $R_P(u, v) = \{u, v\}$. Also $R_P(u, u) = \{u\}$. For $u \in R_P(u, v)$ and $u \neq v$, clearly $v \notin R_P(u, u) = \{u\}$.

Case 2. Suppose $u \leq v$, then by definition $R_P(u, v) = \{z/u \leq z \leq v\}$. If $x \in R_P(u, v)$, then $x \leq v$ and if $v \in R_P(u, x)$, then $v \leq x$, hence $x = v$. Therefore if $x \in R_P(u, v)$ and $x \neq v$ it follows $v \notin R_P(u, x)$.

To show that R satisfies $(b2)$, we consider two cases.

Case 1. Suppose u and v are incomparable. Then $R_P(u, v) = \{u, v\}$, $R_P(u, u) = \{u\}$ and $R_P(v, v) = \{v\}$. Clearly $R_P(u, u) \subseteq R_P(u, v)$ and $R_P(v, v) \subseteq R_P(u, v)$.

Case 2. Suppose $u \leq v$. Let $x \in R_P(u, v)$, then by definition $u \leq x \leq v$. Now if $z \in R_P(u, x)$ then $u \leq z \leq x$. Therefore $u \leq z \leq x \leq v$ and $z \in R_P(u, v)$. Hence $R_P(u, x) \subseteq R_P(u, v)$. Thus R_P is a betweenness on P.

(ii) Consider $R_S(u, v)$ and $u, v \in S$. To show that R_S satisfies $(b1)$, we consider two cases.

Case 1. Suppose u and v are comparable and $u \neq v$. That is $u \leq v$ or $v \leq u$. If $u \leq v$ then $u \wedge v = u$ and by definition $R_S(u, v) = \{z/u \leq z \leq v\}$. Consider $x \in R_S(u, v)$ and $x \neq v$. Suppose $v \in R_S(u, x)$, then either $u \wedge x \leq v \leq u$ or $u \wedge x \leq v \leq x$. From $u \wedge x \leq v \leq u$, it follows $u = v$, which is against the assumption that $u \neq v$. Now consider $u \wedge x \leq v \leq x$. Also we have $u \leq x \leq v$. Both together give $x = v$, a contradiction. Therefore $v \notin R_S(u, x)$. Similar is the case when $v \leq u$. When $u = v$, there is nothing to prove.

Case 2. Suppose u and v are incomparable. Let $x \in R_S(u, v)$ and $x \neq v$. Suppose $v \in R_S(u, x)$, then $u \wedge x \leq v \leq u$ or $u \wedge x \leq v \leq x$. Since u and v are incomparable $u \wedge x \leq v \leq u$ is not possible. Hence the following cases remain.

Case 2.1. Suppose $u \wedge v \leq x \leq u$ and $u \wedge x \leq v \leq x$, then $u \wedge x \leq v \leq x \leq u$. Therefore $v \leq u$, a contradiction.

Case 2.2. Suppose $u \wedge v \leq x \leq v$ and $u \wedge x \leq v \leq x$, then $x = v$, a contradiction. Therefore $v \notin R_S(u, x)$. Hence R_S satisfies $(b1)$. Suppose $x \in R_S(u, v)$ and $z \in R_S(u, x)$. To show that R_S satisfies $(b2)$, we consider four cases.

Case 1. Suppose $u \wedge v \leq x \leq u$, then $u \wedge (u \wedge v) \leq u \wedge x \leq u \wedge u$. Therefore $u \wedge v \leq u \wedge x \leq u$. Together with $u \wedge x \leq z \leq u$ we get $u \wedge v \leq z \leq u$.

Case 2. Suppose $u \wedge v \leq x \leq u$, then as in Case 1, $u \wedge v \leq u \wedge x \leq u$. Together with $u \wedge x \leq z \leq x$, we get $u \wedge v \leq z \leq u$.

Case 3. Suppose $u \wedge v \leq x \leq v$, then $u \wedge (u \wedge v) \leq u \wedge x \leq u \wedge v$. Therefore $u \wedge x = u \wedge v$. Hence from $u \wedge x \leq z \leq u$, we get $u \wedge v \leq z \leq u$.

Case 4. Suppose $u \wedge v \leq x \leq v$, then as in case 3 we have $u \wedge x = u \wedge v$. Hence from $u \wedge x \leq z \leq x$, we get $u \wedge v \leq z \leq v$.

Thus in all four cases $z \in R_S(u, v)$. Therefore R_S satisfies $(b2)$. □

4. Characterization of betweenness of the lattice transit function R_L

Definition 2. *For a lattice L, the lattice transit function* $R_L : V \times V \to 2^V$ *is defined as* $R_L(x, y) = \{z | x \wedge y \leq z \leq x \vee y\}$.

The next two examples show that R_L does not necessarily satisfy axiom $(b1)$.

Example 1. Consider the lattice L_1. See Figure 1(a). Here $R_L(a, b) = \{z | a \wedge b \leq z \leq a \vee b\} = \{z | e \leq z \leq d\} = \{a, b, c, d, e\}$. Now $c \in R_L(a, b)$ and $c \neq b$. Then $R_L(a, c) = \{z | a \wedge c \leq z \leq a \vee c\} = \{z | e \leq z \leq d\} = \{a, b, c, d, e\} = R_L(a, b)$. Therefore $b \in R_L(a, c)$. Thus if $c \in R_L(a, b)$ and $c \neq b$ then $b \in R_L(a, c)$. Therefore L_1 does not satisfy $(b1)$.

Example 2. Consider the lattice L_2. See Figure 1(b). Here $R_L(x, y) = \{z | x \wedge y \leq z \leq x \vee y\} = \{z | u \leq z \leq w\} = \{x, y, u, v, w\}$. Now $v \in R_L(x, y)$ and $v \neq y$. Then $R_L(x, v) = \{z | x \wedge v \leq z \leq x \vee v\} = \{z | u \leq z \leq w\} = \{x, y, u, v, w\} = R_L(x, y)$. Therefore $y \in R_L(x, v)$. Thus if $v \in R_L(x, y)$ and $v \neq y$, then $y \in R_L(x, v)$. Therefore L_2 does not satisfy $(b1)$.

Lemma 1. R_L *satisfies* (b_2) *on any lattice L.*

Proof. Let $z \in R_L(u, x)$, then $u \wedge x \leq z \leq u \vee x \ldots (1)$. Since $x \in R_L(u, v)$, $u \wedge v \leq x \leq u \vee v$. Then $u \wedge (u \wedge v) \leq u \wedge x \leq u \wedge (u \vee v)$ so that $u \wedge v \leq u \wedge x \leq u$. Also $u \vee (u \wedge v) \leq u \vee x \leq u \vee (u \vee v)$. Therefore $u \leq u \vee x \leq u \vee v$. Thus $u \wedge v \leq u \wedge x \leq u \leq u \vee x \leq u \vee v$. Together with (1) we get $u \wedge v \leq z \leq u \vee v$. Therefore $z \in R_L(u, v)$. □

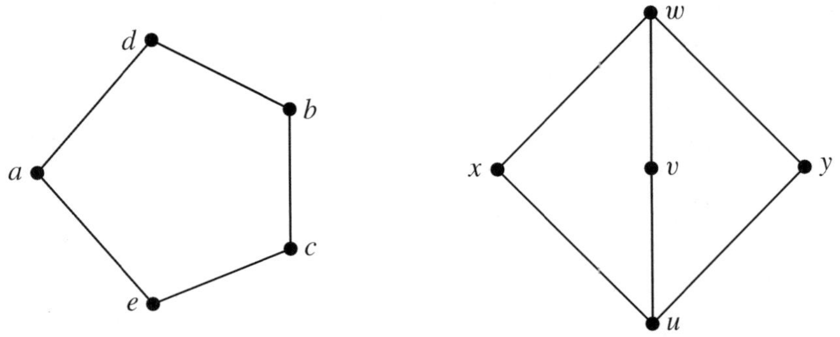

(a) *Pentagonal lattice* L_1 (b) *Diamond lattice* L_2

Figure 1.

We can characterize the betweenness of R_L by forbidden lattices L_1 and L_2.

Theorem 1. *R_L is a betweenness if and only if L is free from L_1 and L_2.*

Proof. By lemma 1, R_L always satisfies $(b2)$. So it suffices to prove that R_L satisfies (b_1) if and only if L is free from L_1 and L_2. Examples 1 and 2 show R_L does not satisfy $(b1)$ if lattice L is not free from L_1 and L_2. So it remains to prove that if L is free from L_1 and L_2 then R_L satisfies $(b1)$. Suppose on the contrary that L is free from L_1 and L_2 and there exists $u, v, x \in L$ with $x \in R_L(u, v)$ and $x \neq v$ satisfying $v \in R_L(u, x)$. First we claim that in this case u and v are incomparable. Suppose on the contrary, they are comparable, then $u \leq v$ or $v \leq u$. Consider the case when $u \leq v$. From $x \in R_L(u, v)$, it follows by definition, $u \wedge v \leq x \leq u \vee v$ and finally $u \leq x \leq v$ since $u \leq v$. Similarly from $v \in R_L(u, x)$, we get $u \wedge x \leq v \leq u \vee x$ and finally $u \leq v \leq x$ since $u \leq x$. Thus we get $v = x$, a contradiction. Similar is the case when $v \leq u$.

Consider $x \in R_L(u, v)$. Then by definition $u \wedge v \leq x \leq u \vee v$ (1)

Therefore $u \wedge (u \wedge v) \leq u \wedge x \leq u \wedge (u \vee v)$ and $u \wedge v \leq u \wedge x \leq u$ (2)

From (1), $u \vee (u \wedge v) \leq u \vee x \leq u \vee (u \vee v)$ and $u \leq u \vee x \leq u \vee v$ (3)

Also from (1), $v \wedge (u \wedge v) \leq v \wedge x \leq v \wedge (u \vee v)$ and $u \wedge v \leq v \wedge x \leq v$ (4)

Again from (1), $v \vee (u \wedge v) \leq v \vee x \leq v \vee (u \vee v)$ and $v \leq v \vee x \leq u \vee v$ (5)

Next we consider two cases.

Case 1. Suppose first x is incomparable with both u and v. In general we then have situation depicted in Figure $2(a)$. From $v \in R_L(u, x)$, it follows by definition $u \wedge x \leq v \leq u \vee x$. Therefore

$$u \wedge (u \wedge x) \leq u \wedge v \leq u \wedge (u \vee x) \text{ and } u \wedge x \leq u \wedge v \leq u$$

Now (2) and (6) together give $u \wedge x = u \wedge v$. Also, from (6), we get

$$u \vee (u \wedge x) \leq u \vee v \leq u \vee (u \vee x). \text{ Therefore } u \leq u \vee v \leq u \vee x$$

Now (3) and (7) together give $u \vee x \leq u \vee v$. Consequently we have situation depicted in Figure 2(b) which does not satisfy b_1 since in that case $x \in R_L(u, v)$ and $x \neq v$ but $v \in R_L(u, x)$. Clearly the lattice shown in Figure 2(b) has the sublattice formed by u, v, $u \wedge v$, $u \vee v$ and $v \vee x$ isomorphic to that of L_1.

Next we consider particular cases that can be derived from the situation depicted in figure 2(b).

Case 1.1. Suppose $v \vee x = u \vee v$. Then we have situation depicted in Figure 2(c) and it contains the sublattice formed by u, v, $u \wedge v$, $u \vee v$ and $v \wedge x$ isomorphic to that of L_1.

Case 1.2. Suppose $v \wedge x = u \wedge v$, then we have the situation depicted in Figure 2(d). Clearly this also contains the sublattice formed by u, v, $u \wedge v$, $u \vee v$ and $v \vee x$ isomorphic to that of L_1.

Case 1.3. Suppose $v \vee x = u \vee v$ and $v \wedge x = u \wedge v$. Then we have situation depicted in Figure 2(e) and it contains the sublattice formed by u, v, x, $u \wedge v$ and $u \vee v$ isomorphic to that of L_2.

Note that the case $v \vee x = v$ or $v \wedge x = v$ is not possible since then x and v would be comparable, a contradiction with the above assumption.

Case 2. Suppose x is comparable with u or v.

Case 2.1. Suppose x is comparable with u.

Suppose first $x \leq u$, then $u \wedge x = x$, $u \vee x = u$. From (2), we get $u \wedge v \leq x \leq u$ and we have the situation depicted in Figure 2(f). From $x \leq u$, it follows $v \wedge x \leq u \wedge v$. Together with (4), we have $u \wedge v = v \wedge x$ and the situation depicted in Figure 2(g). Suppose $v \in R_L(u, x)$, then by definition $u \wedge x \leq v \leq u \vee x$. Therefore $x \leq v \leq u$ as $x \leq u$. Thus $v \leq u$, a contradiction with the fact that u and v are incomparable. Similar is the case when $u \leq x$. Hence the case x is comparable with u is not possible.

Case 2.2. Suppose x is comparable with v.

Suppose first $x \leq v$, then $v \wedge x = x$, $v \vee x = v$. From (4) we get $u \wedge v \leq x \leq v$ so that we have the situation depicted in Figure 2(h). From $x \leq v$, it follows $u \wedge x \leq u \wedge v$. Together with (2) we have $u \wedge x = u \wedge v$ and the situation depicted in Figure 2(i). Suppose $v \in R_L(u, x)$, then by definition $u \wedge x \leq v \leq u \vee x$. Therefore $u \wedge v \leq v \leq u \vee x$ and we have the situation depicted in Figure 2(j). From $u \vee x \leq u \vee v$ and $u, v \leq u \vee x$ it follows $u \vee x = u \vee v$ and we have situation depicted in Figure 2(k) which does not satisfy $(b1)$ since in that case $x \in R_L(u, v)$ and $x \neq v$ but $v \in R_L(u, x)$. Clearly the lattice shown in Figure 2(k) contains the sublattice formed by u, v, $u \wedge v$, $u \vee v$ and x isomorphic to L_1. Similar is the case when $v \leq x$. Hence in all cases there exists a lattice which is isomorphic to L_1 or L_2 which is a contradiction with our assumption. Therefore if L is free from L_1 and L_2, then R_L satisfies axiom $(b1)$ and the proof is complete. □

Since by [16, Theorem 1, pp. 80], a lattice is distributive if and only if it has no sublattice isomorphic to L_1 or L_2, we obtain the following corollary.

Corollary 1. *A lattice L is distributive if and only if the lattice transit function R_L defined on L is a betweenness.*

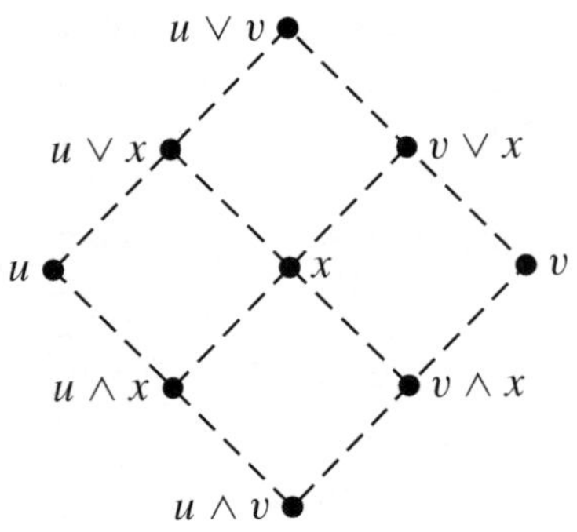

(*a*) *x is incomparable with u and v*

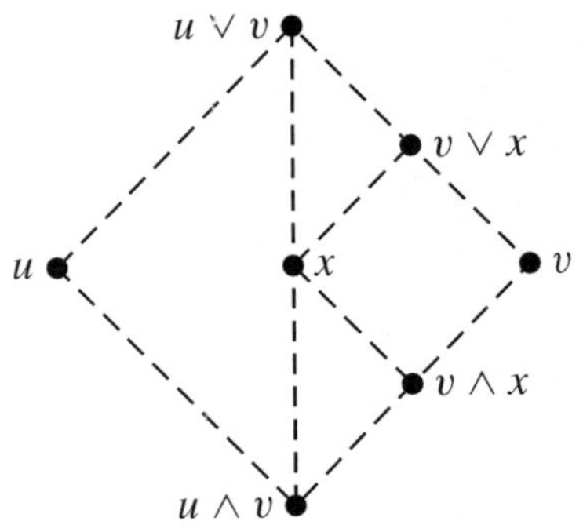

(*b*) $u \wedge x = u \wedge v$ *and* $u \vee x = u \vee v$

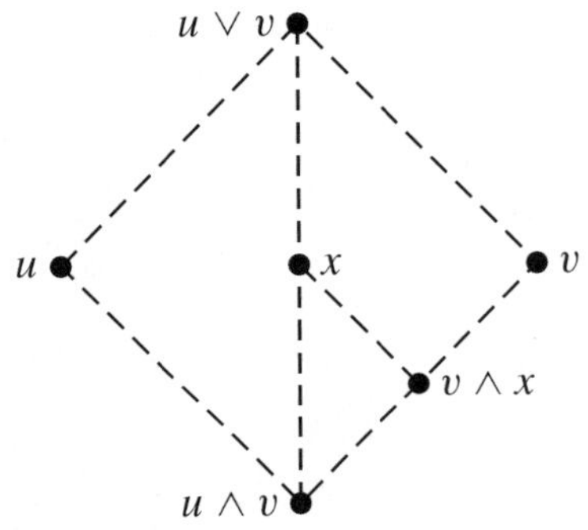

(*c*) $v \vee x = u \vee v$

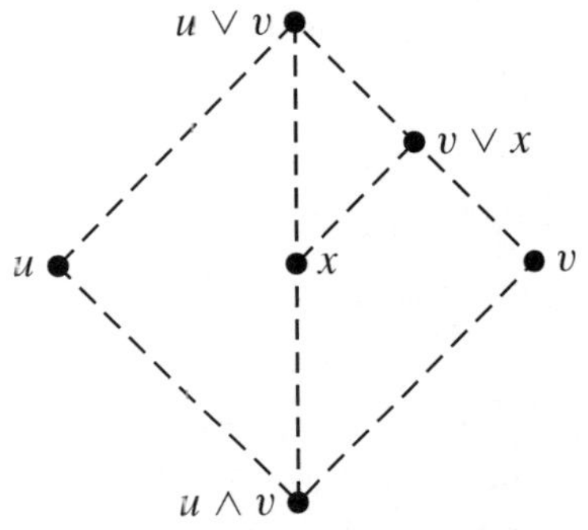

(*d*) $v \wedge x = u \wedge v$

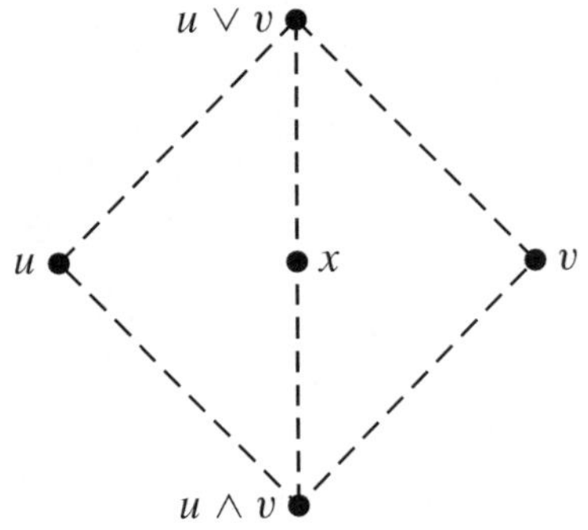

(*e*) $v \vee x = u \vee v$ *and* $v \wedge x = u \wedge v$

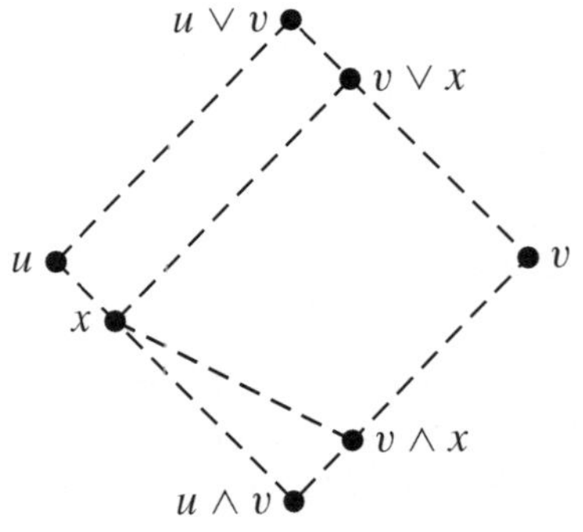

(*f*) $x \leq u$

5. Monotonicity of transit functions R_P, R_S and R_L

We have the following proposition.

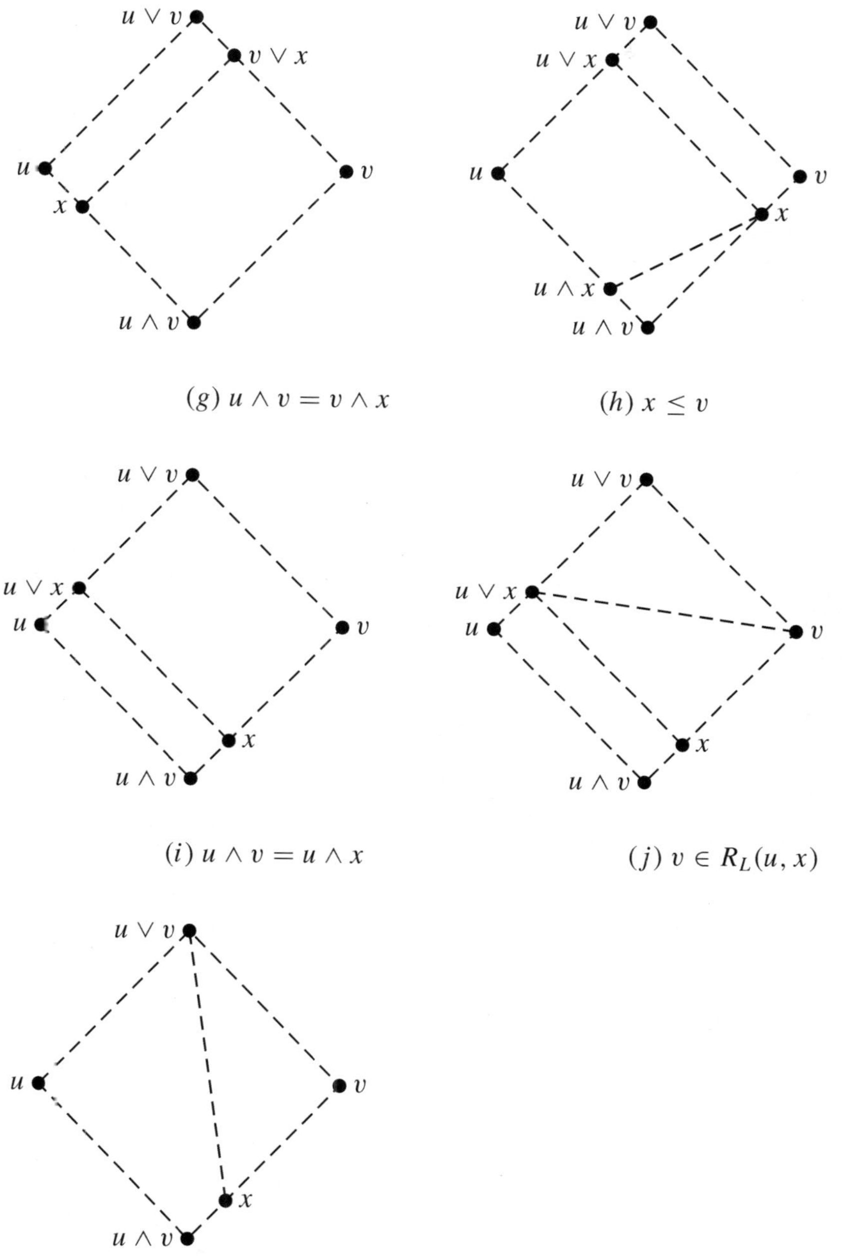

$(g)\ u \wedge v = v \wedge x$

$(h)\ x \leq v$

$(i)\ u \wedge v = u \wedge x$

$(j)\ v \in R_L(u, x)$

$(k)\ u \vee x = u \vee v$

Figure 2.

Proposition 2.

(i) R_P *is monotone on* P.
(ii) R_S *is monotone on* S.
(iii) R_L *is monotone on* L.

Proof.

(i) Consider $R_P(u, v)$ where $u, v \in V$. Suppose first u, v are incomparable. Then by definition $R_P(u, v) = \{u, v\}$ and we are done. Suppose next $u \leq v$. Let

$$x, y \in R_P(u, v) \text{ then } u \leq x, y \leq v \tag{1}$$

Suppose first x, y are incomparable, then $R_P(x, y) = \{x, y\} \subseteq R_P(u, v)$. Suppose next $x \leq y$, then $R_P(x, y) = \{z/x \leq z \leq y\}$. Let $w \in R_P(x, y)$ then $u \leq x \leq w \leq y \leq v$ by (1), therefore $u \leq w \leq v$ and $w \in R_P(u, v)$. Thus $R_P(x, y) \subseteq R_P(u, v)$.

(ii) As the cases meet semilattice and join semilattice are similar, let us consider only the first. Consider $x, y \in R_S(u, v)$, then $u \wedge v \leq x \leq u$ or $u \wedge v \leq x \leq v$ and $u \wedge v \leq y \leq u$ or $u \wedge v \leq y \leq v$. Let $z \in R_S(x, y)$ then $x \wedge y \leq z \leq x$ or $x \wedge y \leq z \leq y$. Since x and y are arbitrary elements of $R_S(u, v)$, u and v are arbitrary elements of S and R_S is $(t2)$, it suffices to consider the following case only.
Consider $u \wedge v \leq x \leq u$, $u \wedge v \leq y \leq u$ and $x \wedge y \leq z \leq x$. Then $u \wedge v \leq x \wedge y \leq u$. Therefore $u \wedge v \leq x \wedge y \leq z \leq x \leq u$. Therefore $u \wedge v \leq z \leq u$. Hence $z \in R_S(u, v)$.
Thus $R_S(x, y) \subseteq R_S(u, v)$. Therefore R_S is monotone.

(iii) Let $x, y \in R_L(a, b)$. Consider $z \in R_L(x, y)$, then

$$x \wedge y \leq z \leq x \vee y \tag{1}$$

Consider $x, y \in R_L(a, b)$, then $a \wedge b \leq x$, $a \wedge b \leq y$. Then

$$a \wedge b \leq x \wedge y \leq x, a \wedge b \leq x \wedge y \leq y. \text{ Therefore } a \wedge b \leq x \wedge y \leq x \vee y \tag{2}$$

Since $x \leq a \vee b$, $y \leq a \vee b$, we have $x \vee y \leq a \vee b$. From (2), $a \wedge b \leq x \wedge y \leq x \vee y \leq a \vee b$. Now by (1), $z \in R_L(a, b)$. Thus $R_L(x, y) \subseteq R_L(a, b)$ for all $x, y \in R_L(a, b)$. □

6. Relation between the underlying graphs of R_P, R_S and R_L and the corresponding covering graphs of P, S and L

We can see that the underlying graph of the semilattice (lattice) transit function is isomorphic to the covering graph of the semilattice (lattice) whereas this may not be the case regarding the underlying graph of the poset transit function and the covering graph of the poset. We summarize these ideas in the following proposition.

Proposition 3.

(i) *The underlying graph G_{R_P} of the poset transit function R_P is isomorphic to the covering graph $G(P)$ of P if and only if P is a chain.*

(ii) *The underlying graph G_{R_S} (G_{R_L}) of the semilattice (lattice) transit function R_S (R_L) is isomorphic to the covering graph $G(S)$ ($G(L)$) of the semilattice S (lattice L).*

Proof.

(i) Suppose P is a chain. Take $u, v \in V$. Suppose u is adjacent to v in G_{R_P}. Then $|R_P(u, v)| = 2$. Hence either v covers u or u covers v in P since P is a chain. Therefore u is adjacent to v in the covering graph $G(P)$ of P. Now we find that adjacency in the underlying graph G_{R_P} follows from the adjacency in the covering graph $G(P)$ of P without using the chain condition. Hence G_{R_P} is isomorphic to the covering graph $G(P)$ of P.
Conversely let $u, v \in V$. Suppose u and v are incomparable in P. Then they are nonadjacent in the covering graph $G(P)$ of P. But by definition, $R_P(u, v) = \{u, v\}$. Therefore they are adjacent in G_{R_P}. Hence in the case when P is not a chain the covering graph $G(P)$ of P and the underlying graph G_{R_P} are not isomorphic.

(ii) From the definitions it follows that the underlying graph G_{R_S} (G_{R_L}) of R_S (R_L) is isomorphic to the covering graph $G(S)$ $(G(L))$ of S (L). □

7. Conclusion

In graphs, the theory of transit functions is now a developing branch as can be understood from the references. Even though there is some literature on the interval convexities in posets, lattices and semilattices, this is not the case on the betweenness and other properties of transit functions in these important discrete structures. This paper is an attempt to study the betweenness property of the natural transit functions on posets, lattices and semilattices using the property of their underlying structure. Geometricity is an important notion which is not explored much in the case of transit functions on posets. Nevertheless Van de vel [25] has discussed two special types of geometric transit functions on posets and lattices. Mulder in his paper, "Transit functions on graphs (and posets)" [20] which appears in this volume has proposed several interesting problems on transit functions on graphs. He also has mentioned that transfer of ideas between one transit function to another is interesting in certain cases. So we feel that this transfer of ideas from transit functions on graphs to transit functions on posets will be interesting and relevant.

Acknowledgement

The authors thank the referee for his valuable remarks and suggestions. The authors are obliged much to the precious guidance and advice made by Dr. Manoj Changat, Department of Futures Studies, University of Kerala, Thiruvananthapuram.

References

[1] H.-J. Bandelt and V. Chepoi, Metric graph theory and geometry, a survey, *Contemp. Math.*, to appear.
[2] H.-J. Bandelt and V. Chepoi, The algebra of metric betweenness I: Subdirect representation and retraction, *European J. Combin.*, in press.

[3] H.-J. Bandelt and V. Chepoi, The algebra of metric betweenness II: Axiomatics of weakly median graphs, *European J. Combin.*, to appear.

[4] H.-J. Bandelt and H. M. Mulder, Distance – hereditary graphs, *J. Combin. Th. Ser. B*, **41** (1986) 182–208.

[5] H.-J. Bandelt and H. M. Mulder, Pseudo – modular graphs, *Discrete Math.*, **62** (1986) 245–260.

[6] H.-J. Bandelt, H. M. Mulder and E. Wilkeit, Quasi-median graphs and algebras, *J. Graph Theory*, **18** (1994) 681–703.

[7] J. Calder, Some elementary properties of interval convexities, *J. London Math. Soc.*, **3** (1971) 422–428.

[8] M. Changat, S. Klavžar and H. M. Mulder, The all path transit function of a graph, *Czech. Math. J.*, **51** (2001) no. 126, 439–448.

[9] M. Changat and J. Mathew, Induced path transit function, monotone and Peano axioms, *Discrete Math.*, **286** (2004) 185–194.

[10] M. Changat and J. Mathew, Interval monotone graphs: minimal path convexity, Proceedings of the conference on Graph connections, New Delhi (1999) 87–90.

[11] M. Changat, H. M. Mulder and G. Sierksma, Convexities related to path properties on graphs, *Discrete Math.*, **290** (2005) 117–131.

[12] V. Chepoi, Isometric subgraphs of Hamming graphs and *d*-convexity, *Cybernetics*, **24** (1988) 6–10.

[13] V. Chepoi, Classifying graphs by metric triangles, *Metody Diskretnogo Analiza*, **49** (1989) 75–93 (in Russian).

[14] V. Chepoi, Convexity and local conditions on graphs, Studies in Applied Mathematics and Information Science, pp. 184–191, Stiinta Chisinau (1990) (in Russian).

[15] G. Birkhoff, Lattice Theory, AMS Coll. Publ., Providence (1967).

[16] G. Grätzer, General Lattice Theory, Second Edition, Birkhauser Verlag, Basel (1998).

[17] W. Imrich and S. Klavžar, Product Graphs: Structure and Recognition, Wiley Interscience Publication, New York (2000).

[18] R. E. Jamison, Tietze's convexity theorem for semilattices and lattices, *Semigroup Forum*, **15** (1978) 357–373.

[19] M. A. Morgana and H. M. Mulder, The induced path convexity, betweenness and svelte graphs, *Discrete Math.*, **254** (2002) 349–370.

[20] H. M. Mulder, Transit functions on graphs (and posets) (in this volume) 117–130.

[21] H. M. Mulder, The Interval Function of a Graph, Mathematisch Centrum, Amsterdam (1980).

[22] L. Nebesky, A characterization of the interval function of a connected graph, *Chech. Math. J.*, **44** (1994) no. 119, 173–178.

[23] M. Sholander, Trees, lattices, order and betweenness, *Proc. Amer. Math. Soc.*, **3** (1952) 369–381.

[24] J. C. Varlet, Remark on distributive Lattices, *Bull. Acad. Polon. Sci. Sŕ. Sci. Math. Astronom. Phys.*, **23** (1975) 1143–1147.

[25] M. van de Vel, Theory of Convex Structures, North-Holland, Amsterdam (1993).

[26] M. van de Vel, Binary convexities and distributive lattices, *Proc. London Math. Soc. (3)*, **48** (1984) 1–33.

[27] M. van de Vel, Dimension of binary convex structures, *Proc. London Math. Soc. (3)*, **48** (1984) 34–54.

[28] M. van de Vel, Lattices and semilattices: a convex point of view, in: Continuous Lattices and Their Applications, *Lecture Notes Pure Appl. Math.*, Dekker, New York, **101** (1984) 279–302.

Joint Proc. Int. Wrk.shps. Conv. in Discr. Struct. and Metric & Conv. Graph Th. No. 5, 2008, pp. 117–130.

Transit Functions on Graphs (and Posets)

Henry Martyn Mulder

Econometrisch Instituut, Erasmus Universiteit, P.O. Box 1738, 3000 DR Rotterdam, The Netherlands
e-mail: hmmulder@few.eur.nl

Abstract. The notion of transit function is introduced to present a unifying approach for results and ideas on intervals, convexities and betweenness in graphs and posets. Prime examples of such transit functions are the interval function I and the induced path function J of a connected graph. Another transit function is the all-paths function. New transit functions are introduced, such as the cut-vertex transit function and the longest path function. The main idea of transit functions is that of 'transferring' problems and ideas of one transit function to the other. For instance, a result on the interval function I might suggest similar problems for the induced path function J. Examples are given of how fruitful this transfer can be. A list of Prototype Problems and Questions for this transferring process is given, which suggests many new questions and open problems.

Keywords. Transit function, interval function, induced path, convexity, betweenness, path function, block graph.

Subject Classification: 05C99, 05C12, 05C38, 05C75

1. Introduction

The geodetic interval function I and the geodetic convexity of a connected graph are important tools in problems that involve distance and shortest paths (i.e. geodesics). But there are many other ways how one would like to move around in a graph. Another prime example is that where one uses induced paths instead of shortest paths to move from one vertex to another. This yields another type of convexity on a graph. As can be expected, proof techniques for the geodetic case and the induced path case are very different. On the other hand, problems and results that are obtained for the interval function I might suggest problems and questions for the induced path function J. An interesting example of this approach is the following. An interval $I(u, v)$ in a connected graph consists of all vertices that are in between u and v, that is, lie on some shortest u, v-path. A median graph is defined by the fact that $|I(u, v) \cap I(v, w) \cap I(w, u)| = 1$, for any three vertices u, v, w. These graphs allow a rich structure theory, and there are by now many applications available in such diverse areas as evolutionary biology, chemistry, location theory, consensus theory, and so on and so forth (see e.g. [19,16,14]). This condition on the intersection of three intervals and the theory that followed from it suggested to me already in 1991 to consider the question what one might say about graphs for which there are similar conditions on the intersection $J(u, v) \cap J(v, w) \cap J(w, u)$, for any three vertices u, v, w. Here $J(u, v)$ is the set of all vertices on induced u, v-paths.

Because of the big difference between shortest path and induced path one could not expect right away that this idea would lead to anything. But interestingly enough, non-trivial results could be obtained, see [17], where results as well as proof techniques are quite different. In January 1998 at the Conference on "Graph Connections" at Cochin University of Science and Technology the idea came to me that this transferring of problems from one such function to another could be utilized to generate many problems and questions that might be worthwhile to study in their own right. This is the underlying motivation for the idea of transit function. A *transit function* is a very general notion that unifies the many ways how one could move around in a graph. But it can also be used to study other structural aspects of graphs, such as the structure of the blocks and cut-vertices, or convexity and betweenness. Moreover, other discrete structures than graphs, for instance partially ordered sets, or hypergraphs, can be considered as well. The general notion of transit function in itself might not be very interesting. But for this transferring process of ideas, problems and questions from one transit function to another it has already proven to be worthwhile, as can be concluded from the examples and references below.

A transit function $R : V \times V \to 2^V$ satisfies three simple axioms. The basic idea is: given a set V, a discrete structure σ on V and a transit function R on V, study the interrelations between σ and R. In our case, the structure σ is the set of edges E of a graph. The other pertinent structure from our point of view is that of a partial order $\leq$ on V. We will discuss this case only in a brief manner. First the notion of transit function on a discrete structure (V, σ) is introduced in its most general form, with associated notions, amongst which 'interval', 'convexity', and 'betweenness'. In this framework, twenty prototype questions and problems are formulated. Most of these were suggested by known results on the interval function or the induced path function. Then we give examples, from the literature, but also some new ones to show the fruitfulness of this approach. Some ideas below are just first thoughts on the topic and may need adjustment or further formalization before they can be properly pursued.

2. Transit functions

□ A *discrete structure* (V, σ) consists of a finite set V and a 'structure' σ on V. Prime examples of discrete structures on V are a graph $G = (V, E)$, where $\sigma = E$ is the edge set of the graph, and a partially ordered set, poset for short, $(V, \leq)$, where $\sigma = \leq$ is a partial ordering on V. We denote the power set of V by 2^V.

□ A *transit function* on a discrete structure (V, σ) is a function $R : V \times V \to 2^V$ satisfying the three transit axioms

$$(t1)\quad u \in R(u, v),\ \text{ for any } u \text{ and } v \text{ in } V,$$

$$(t2)\quad R(u, v) = R(v, u),\ \text{ for all } u \text{ and } v \text{ in } V,$$

$$(t3)\quad R(u, u) = \{u\},\ \text{ for all } u \text{ in } V.$$

Axioms as these, which are phrased in terms of the function only, are called *transit axioms*. Axiom $(t3)$ is to avoid degenerate cases. One might also consider functions

R that do not satisfy $(t3)$. These could then be called *weak transit functions*. Note that the structure σ is not involved in the definition of a transit function. But, of course, the focus of our interest will be on the interplay between the function R and the structure σ. Usually, R will just be defined in terms of σ.

In this paper our main point of focus is transit functions on graphs. Then we are interested in the interrelations between the two structures E and R on V. Basically, a transit function on a graph G describes how we can get from vertex u to vertex v: via vertices in $R(u, v)$. A prime example is the *interval function* I defined by

$$I(u, v) = \{w | d(u, w) + d(w, v) = d(u, v)\},$$

see [19] for an extensive study of the interval function. In the terminology of this paper one would call the interval function the *geodetic transit function*.

□ The *trivial transit function* $\mathbf{1}$ on (V, σ) is defined by $\mathbf{1}(u, u) = \{u\}$, for all u in V and $\mathbf{1}(u, v) = V$, for all distinct u, v in V. The *discrete transit function* $\mathbf{0}$ is defined by $\mathbf{0}(u, u) = \{u\}$, for all u in V, and $\mathbf{0}(u, v) = \{u, v\}$, for all distinct u, v in V.

□ Let R and S be two transit functions on (V, σ). The *join* of R and S is the transit function $R \vee S$ defined by $R \vee S(u, v) = R(u, v) \cup S(u, v)$. The *meet* of R and S is the function $R \wedge S$ defined by $R \wedge S(u, v) = R(u, v) \cap S(u, v)$. With this join and meet the family of transit functions on (V, σ) is a lattice with the discrete transit function $\mathbf{0}$ as universal lower bound and the trivial transit function $\mathbf{1}$ as universal upper bound.

□ The *underlying graph* G_R of transit function R is the graph with vertex set V, where distinct u and v in V are joined by an edge if and only if $|R(u, v)| = 2$.

□ The *transit hypergraph* of R has V as vertex set, and, for distinct u and v, a set $R(u, v)$ is an edge if and only if for each w in $R(u, v)$ distinct from u, v we have $R(u, v) = R(u, w)$. The *transit graph* of R is the underlying graph of the transit hypergraph, that is, u and v are adjacent if they are in the same edge of the transit hypergraph. Note that there may be other types of hypergraph to be associated with a specific transit function.

□ A set $W \subseteq V$ is an *R-convex set* in (V, σ) if $R(u, v) \subseteq W$, for all u, v in W. The family $\mathcal{C}_R$ of R-convex sets in (V, σ) is an abstract convexity: it is closed under intersections and both $\emptyset$ and V are R-convex sets. The family of R-convex sets is called the *R-convexity* on G.

Note that axiom $(t3)$ implies that the R-convexity satisfies the so-called separation axiom S_1, that is, singletons are R-convex. Also note that in the infinite case there is yet another condition to be satisfied by an abstract convexity: it must be closed under nested unions. Let us keep in mind that, in case we encounter a convexity that is not S_1, we may drop axiom $(t3)$ to obtain a weak transit function R to be associated with this convexity.

An R-convexity of a transit function R is also known as an *interval space*, cf. [34], and may be studied from the viewpoint of abstract convexity theory. Thus one might ask for the usual convexity parameters of a specific R-convexity. In the context of transit functions we will focus mainly on other types of questions: those related to the basic ideas of transit functions on discrete structures as exemplified below.

□ Let $\mathcal{C}$ be any convexity on the finite set V, that is, $\mathcal{C}$ is a family of subsets of V closed under intersection and containing $\emptyset$ and V. The sets in $\mathcal{C}$ are the convex sets of the convexity. The convex hull $co_{\mathcal{C}}(W)$ of a subset W of V with respect to the convexity $\mathcal{C}$ is the smallest convex set containing W, that is, the intersection of all convex sets containing W. We assume that the convexity $\mathcal{C}$ is S_1, so that all singletons are convex. The transit function $R_{\mathcal{C}}$ defined by $R_{\mathcal{C}}(u, v) = co_{\mathcal{C}}(u, v)$ is the *canonical transit function* of the convexity. Note that the convexity defined by $R_{\mathcal{C}}$ need not be $\mathcal{C}$ itself. For instance, take the convexity $\mathcal{C}$ consisting of V and all subsets of V of size at most 2. Then the convexity defined by $R_{\mathcal{C}}$ consists of all subsets of V.

□ Let R be a transit function on (V, σ). Let $\mathcal{C} = \mathcal{C}_R$ be the R-convexity, and let R^* be the canonical transit function of $\mathcal{C}$. Then, clearly, R and R^* define the same convexity $\mathcal{C}$ on V, even if R and R^* do not coincide. So, unless $R = R^*$, trivially there is always another transit function defining the same convexity as R. If there is no transit function S on (V, σ) distinct from both R and R^* such that the R-convexity and the S-convexity coincide, then we say that the R-convexity is *uniquely determined by* R.

□ For a transit function R, we define

$$R(u, v, w) = R(u, v) \cap R(v, w) \cap R(w, u).$$

This intersection with respect to the interval function I of a connected graph plays a major role in the study of median graphs and modular graphs, cf. [19,14], see also Section 4.1 below.

□ The discrete structure (V, σ) is said to be *R-monotone* if the sets $R(u, v)$ are R-convex, for all u, v in V, cf. the notion of interval monotone graph in [19].

□ For a transit function R, we define the transit function F_R by

$$F_R(u, v) = \{w | R(u, w) \cap R(w, v) = \{w\}\}.$$

In case $R = I$, the transit function F_I played a role in the study of Hamming graphs, i.e. Cartesian products of complete graphs, see [19]. Of course, at that time it was not yet phrased in terms of transit functions.

□ Let R be a transit function on (V, σ), let W be a subset of V, and let z be a vertex of G outside W. A vertex x in W is an *R-gate* for z in W if x lies in $R(z, w)$, for any w in W. The set W is called an *R-gated set*, or an *R-prefiber*, if every vertex z outside W has a unique R-gate in W. We note here that the notions of R-gate and R-gated set may need some adjustments, or may give rise to some specializations. For instance, here we do not (yet?) require that each vertex w inside an R-gated set W has a unique R-gate in W. In the case of the interval function I, each vertex x in a set W trivially is its own gate in W.

The notion of gate was first introduced by Goldman and Witzgall [12] with respect to geodesics, that is, in our current terminology with respect to the interval function I.

It played an essential role in the study of quasi-median graphs, which generalize median graphs to the non-bipartite case, see [19,2]. Tardif [33] proposed the term prefiber for I-gated set, because these sets play an important role in the embedding of a graph into a Cartesian product of graphs. Loosely speaking, one may consider the Cartesian product to be a fabric of graphs woven with fibers. The fibers are the factors in the product. When one wants to embed a graph in a product, one of the main problems is to find suitable factors, or fibers, for this product. Candidates are the I-gated sets, or *prefibers*.

□ For a transit function R, we define the transit function Q_R by

$$Q_R(u,u) = \{u\},$$
$$Q_R(u,v) = \{\, w \in R(u,v) \mid v \notin R(u,w), u \notin R(v,w) \,\}.$$

This transit function highlights the elements w that might be "between" u and v. So it may play a role in the study of betweenness, see the next item.

□ A transit function R on (V, σ) is a *betweenness* if it satisfies the additional *betweenness axioms*:

$(b1)$ if $x \in R(u,v)$ and $x \neq v$, then $v \notin R(u,x)$, for u, v in V,

$(b2)$ if $x \in R(u,v)$, then $R(u,x) \subseteq R(u,v)$, for u, v in V.

Note that, as for transit functions, a betweenness can just be defined on the set V by ignoring the structure σ. Thus one can develop an abstract theory of betweenness as in [31]. The notion of betweenness in the above sense was first introduced in [17]. For a slightly different notion of betweenness, see [30].

□ In the rest of this section we restrict ourselves to graphs, that is, σ is the edge set of a graph $G = (V, E)$. Let Φ be a property of paths. A *Φ-path* is a path having property Φ. Formally, one takes Φ to be a subset of all paths in G, so that a Φ-path is a path in Φ. We will only consider *feasible path properties*, that is, path properties Φ such that for each pair of vertices in G there exists a Φ-path. A *Φ-path transit function*, or *Φ-path function* in short, is a transit function R on G such that

$$R(u,v) = \{\, x \mid x \text{ is on some } \Phi\text{-path between } u \text{ and } v \,\}.$$

A *path transit function*, or *path function* in short, is a Φ-path transit function for some feasible path property Φ. The family of path functions on a graph is a join semilattice. The all-paths function A is the universal upper bound of this semilattice, see Section 4.4. See [10] for a first systematic study of path functions.

□ Let $P = u_1 \to u_2 \to \cdots \to u_k$ be a path in G. Let z_i be a vertex not on P but adjacent to two consecutive vertices u_i, u_{i+1} of P. Then we say that the path $Q = u_1 \to u_2 \to \cdots \to u_i \to z_i \to u_{i+1} \to \cdots \to u_k$ is obtained from P by "replacing" the edge $u_i \to u_{i+1}$ by a triangle. A *triangular extension* of a path P is a path Q obtained from P by replacing some of the edges of P by triangles. We will call P a triangular extension of itself as well. Let Φ be a (feasible) path property on G. Then $\Phi^{\triangle}$ is the path property defined by

$$\Phi^{\triangle} = \{Q | Q \text{ is a triangular extension of some path in } \Phi\}.$$

Note that we have $\Phi \subseteq \Phi^{\triangle}$, with equality if and only if no path in Φ is involved in a triangle. In particular, we have equality in the case of a triangle-free graph. The path property $\Phi^{k\triangle}$ is defined recursively by $\Phi^{0\triangle} = \Phi$ and $\Phi^{k\triangle} = (\Phi^{(k-1)\triangle})^{\triangle}$, for $k \geq 1$. And for $m \geq 1$ with $\Phi^{(m+1)\triangle} = \Phi^{m\triangle}$, we write $\Phi^{m\triangle} = \Phi^{\triangle^*}$. If R is the Φ-path transit function, then we write $R^{k\triangle}$ for the $\Phi^{k\triangle}$-path function, and $R^{\triangle^*}$ for the $\Phi^{\triangle^*}$-transit function.

□ A *rapid transit function* on (V, σ) is a function $r : V \times V \to 2^V$ satisfying the three rapid transit axioms

($rt1$) if $r(u, v) \neq \emptyset$, then $u \in r(u, v)$,

($rt2$) $r(u, v) = r(v, u)$, for all u and v in V,

($rt3$) for any two elements u and v in V, there exists a finite sequence of elements $u_1, u_2, \ldots, u_{k(u,v)}$ such that $u = u_1$ and $v = u_{k(u,v)}$ and $r(u_i, u_{i+1}) \neq \emptyset$, for $i = 1, \ldots, k(u, v)$.

A rapid transit function tells us what direct connections we have: if $r(u, v)$ is nonempty, then we have a direct connection, otherwise we have to change at intermediate elements given by ($rt3$). The underlying graph G_r of the rapid transit function r is defined by: uv is an edge in G_r if and only if u and v are distinct vertices with $r(u, v) \neq \emptyset$. An example of a rapid transit function on a graph $G = (V, E)$ is: $r_k(u, v) = \{u, v\}$ if $d(u, v) \leq k$. The underlying graph of r_k is the k-th power of G. Hence r_1 just gives the edges of the graphs. The theme of rapid transit functions will be pursued elsewhere.

3. Prototype problems and questions

In this section we collect a number of prototype problems and questions. Some were inspired by known results on the interval function and are now presented for any transit function. We will give the references to the pertinent literature in the following sections, where we discuss specific examples of transit functions. Some of the problems might seem a little bit cryptic at first reading, but they will make sense in conjunction with the examples in the following sections.

We formulate all the problems and questions only for transit functions on graphs, but almost all of these can also be formulated for other discrete structures (V, σ).

1. Characterize transit function R solely in terms of transit axioms on R. This problem becomes relevant as soon as the transit function is defined in terms of the structure σ, for instance in the case of path functions.
2. Let $\mathcal{C}$ be a convexity on G, i.e. $\mathcal{C}$ is a family of subsets of V that is closed under intersections and both $\emptyset$ and V are in $\mathcal{C}$. Is there a transit function R on G such that $\mathcal{C}$ is precisely the R-convexity on G?

3. Let R be a transit function on G. Is the R-convexity uniquely determined by R? If not, determine all transit functions S such that the S-convexity and the R-convexity coincide.
4. Let R be a transit function on a graph G. Are G and G_R isomorphic? If not, characterize the graphs G for which G and G_R are isomorphic.
5. Characterize the R-monotone graphs, for transit function R.
6. Let R be a transit function. Study $R(u, v, w)$. For example, is it usually empty, or nonempty, or a large set? For which graphs is $|R(u, v, w)| \leq 1$, or $|R(u, v, w)| \geq 1$, for all triples of (distinct) vertices u, v, w. For which graphs is there always equality?
7. Let R be a transit function. Characterize the graphs for which R is a betweenness.
8. Let R be a transit function. Is there a feasible path property Φ such that R is a Φ-path transit function?
9. Let R be a Φ-path transit function. Study the above problems for R^{Δ}, and $R^{k\Delta}$. What is the smallest k such that $R^{k\Delta} = R^{\Delta^*}$ on G?
10. Let R be a transit function. Study parameters of the R-convexity, such as the Helly number, the Carathéodory number, the Radon number, the Tverberg number, and the exchange number.
11. Let R be a transit function. It is easy to see that, if $|R(u, v, w)| \geq 1$, for all triples of (distinct) vertices u, v, w, then the Helly number of the R-convexity is equal to 2. What can be said about the converse: if the Helly number of the R-convexity is 2, is then necessarily $|R(u, v, w)| \geq 1$, for all triples of (distinct) vertices u, v, w? If not, characterize the graphs with Helly number 2 for which we have $|R(u, v, w)| \geq 1$, for all triples of (distinct) vertices u, v, w.
12. Let R be a transit function on a connected graph G. For which graphs G are the R-convex sets connected in G? What properties should R have such that R-convex implies connected on any connected graph G?
13. Let R be a transit function on a connected graph G. Characterize the R-gated sets. Do the R-gated sets form a convexity on G? And if so, is there a transit function Q such that the R-gated sets form the Q-convexity?
14. Let Φ be a feasible path property on G, and let R be the Φ-path transit function on G. Let W be an R-gated set, let u be a vertex outside W with x as (unique) gate in W, and let w be an arbitrary vertex in W. By definition, we can find a Φ-path between u and w passing through x. What properties should Φ have so that we can find a Φ-path P between u and w such that the subpath of P between x and w is contained in W?
15. Let G be a connected graph. Study the join semilattice of path functions on a graph G. For what graphs is this semilattice a lattice?
16. Let R be a transit function on G. Study the transit graph of R. For which graphs is the transit graph precisely G itself?
17. Let R be a transit function on G. Study the transit function F_R, for special instances of R.
18. Let R be a transit function on G. Study the transit function Q_R, for special instances of R.

19. Let R be a transit function. For which graphs do we have $R = F_R$, or $R = Q_R$?
20. Given a transit function R, does R have any interesting new properties that have not yet been studied for other transit functions? If so, then use these to formulate new Prototype Problems and Questions for transit functions.

4. Examples

In this section we present some well known examples of transit functions on graphs and propose a couple of possibly interesting new ones. The interval function I and the induced path function J are well studied. We just list some of the main results related to the prototype problems and questions above without going into detail. For the all-paths transit function a number of prototype questions have been settled. In this case we include more details to give examples of the type of results that were inspired by the transit function approach. The prefiber transit function has already been used a couple of times in different guises in the literature. But, from the viewpoint of transit functions, it has only been studied recently, see [21]. The Δ-convexity played a role in the study of quasi-median graphs. Here a transit function for this convexity is given, and we prove some first simple results. Many of the prototype problems may be relatively easy ones for this transit function. But it is a transit function that has its uses and is not a path function. The cut-vertex transit function is another example of a non-trivial transit function that is not a path function. We include some first obvious results. The longest path function and the single paths functions are proposed as possibilities to extend the approach of transit functions. Except for the all-paths function and the longest path function, the idea of R^{Δ} and further triangular extensions make sense, but they are not listed separately here.

4.1 *The geodetic transit function*

The *interval function* I of a connected graph G is, of course, *the* prime example of a transit function. It is defined by

$$I(u, v) = \{x \mid x \text{ lies on a shortest } u, v\text{-path}\}.$$

Its origins probably belong to the folklore of graph theory. The first published use of the term might be in [19], which is certainly the first systematic study of the interval function. It was characterized in terms of transit axioms only by Nebeský, [25]. It is easily seen to be a betweenness on any graph, see [19]. The parameters of the I-convexity were determined by Duchet [11]. The I-monotone graphs were characterized in [19] by a forbidden minor. In general the set $I(u, v, w)$ may be empty. The graphs with $I(u, v, w)$ always nonempty are the *modular graphs*. The graphs, for which $|I(u, v, w)| = 1$ for all triples of vertices u, v, w, are known as *median graphs*. These were introduced independently by various authors: Avann [1], Nebeský [23], and Mulder and Schrijver [22]. By now, a rich structure theory is available for these graphs and related structures, see e.g. [19,16,14]. There are also many interesting generalizations of median graphs, to name but one: the quasi-median graphs (see [19,2]).

For the triangular extension I^{Δ} see for instance [10]. Prototype Question 3 above is still open.

Open Problem 4.1.1. Is the geodesic convexity of a connected graph G uniquely determined by its interval function I?

4.2 *The induced path transit function*

The *induced path function*, or *minimal path function* on a graph G is the path transit function $J : V \times V \to 2^V$, where J is defined by

$$J(u, v) = \{x | x \text{ lies on an induced } u, v\text{-path}\}.$$

The J-convexity and its parameters were studied by Duchet [11]. The graphs, for which J is a betweenness, were characterized by forbidden subgraphs in [17]. Usually, $J(u, v, w)$ is a big chunk of vertices. The graphs with $|J(u, v, w)| = 0$ for all triples of distinct vertices u, v, w, and those with $|J(u, v, w)| = 1$ for all triples of distinct vertices u, v, w, were characterized in [17]. The *svelte graphs*, i.e. the graphs with $|J(u, v, w)| \leq 1$ for all triples of vertices u, v, w, were characterized by forbidden subgraphs in [17]. This paper was the first instance of fruitfully transferring questions or ideas from one transit function, viz. I, to another, viz. J, with unexpected new problems that were interesting in their own right. The J-monotone graphs are characterized by Changat and Mathews in [5], see also [7]. The theme of betweenness and monotonicity for J is pursued in [8,7,9]. Another question that arose because of the transit function approach was the problem whether the induced path function could be characterized by transit axioms only. Surprisingly enough this is not possible, as Nebeský [28] has shown using first order logic. Many problems are still open, for instance the following challenging problem.

Open Problem 4.2.1. Is the minimal path convexity of a graph uniquely determined by the minimal path function J?

The J^{Δ}-convexity has been studied by Changat and Mathews, see [4].

4.3 *The* Δ*-convexity*

Let $G = (V, E)$ be a connected graph. A subset W of V is Δ*-convex* if the following holds: if W contains an edge of a triangle, then it contains the whole triangle. Clearly, the family of Δ-convex sets is a convexity on G. In [19] a Δ-convex set was called Δ-closed. The Δ-closed sets played an essential role in the characterization of quasi-median graphs. It is easily seen that the following transit function defines the Δ-convexity:

$$\delta(u, v) = \{u, v\} \text{ if } u \text{ and } v \text{ are not adjacent,}$$

$$\delta(u, v) = I^{\Delta}(u, v) \text{ if } u \text{ and } v \text{ are adjacent.}$$

Thus, for adjacent vertices, $\delta(u, v)$ consists of all the cliques containing the edge uv. We restrict ourselves here to some first simple results. Clearly, δ is a path function if and only if G is complete. The following Lemma is obvious.

Lemma 1. *$\delta(u, v, w) \neq \emptyset$ if and only if u, v, w induce a triangle in G.*

Note that this means that Prototype Problem 6 reduces to a triviality.

Proposition 2. *A graph G is δ-monotone if and only if G does not contain $K_{1,1,2}$ as an induced subgraph.*

Proof. First assume that G contains an induced $K_{1,1,2}$ on the vertices u, v, x, y, where x, y are the nonadjacent vertices. Then all four vertices are in $\delta(u, v)$, whereas y is not in $\delta(u, x)$. So the latter set is not δ-convex.

Next assume that G does not contain an induced $K_{1,1,2}$. Pick any two distinct vertices u, v in G. If u, v are not adjacent, then $\delta(u, v) = \{u, v\}$, which is clearly δ-convex. So assume that u, v are adjacent. Since there is no induced $K_{1,1,2}$ in G, the set $\delta(u, v)$ is a clique K such that there is no triangle sharing exactly two vertices with K. Hence $\delta(u, v) = \delta(x, y)$, for any two distinct vertices x, y in $\delta(u, v)$. So $\delta(u, v)$ is δ-convex. □

4.4 *The all-paths transit function*

The results in this section can be found in [3]. Let G be a connected graph. The *all-paths transit function* A on G is the transit function defined by

$$A(u, v) = \{w | w \text{ lies on some path between } u \text{ and } v\}.$$

Clearly, $A(u, u) = \{u\}$, for any u. Furthermore, for distinct u and v, it follows that $A(u, v)$ consists of all blocks in G intersecting $I(u, v)$ in at least two vertices. Here a *block* either is a maximal 2-connected subgraph or it consists of the two ends of a cut-edge. A set W in G is A-convex if and only if it induces a connected subgraph such that, if W contains two vertices of a block, then it contains the whole block. In particular, $A(u, v) = \{u, v\}$ if and only if it is a block, that is, uv is a cut-edge. Thus the all-paths transit function highlights the block structure of the graph. For instance, the edges in the transit hypergraph of A are precisely the blocks of G.

Theorem 3. *Let G be a connected graph. Then $G = G_A$ if and only if G is a tree.*

Theorem 4. *The A-prefibers are precisely the A-convex sets.*

Theorem 5. *Let G be a graph. Then G is a tree if and only if $|A(u, v, w)| = 1$, for every triple of vertices u, v, w of G.*

The block closure G^* of a connected graph G is the graph obtained from G by joining two vertices whenever they are in the same block of G. Thus G^* is the block graph with the same blocks as G. Recall that a *block graph* is a graph in which each block induces a complete subgraph, whence it is a tree-like structure built from complete graphs.

Theorem 6. *Let G be a connected graph. Then G^* is the transit graph of A.*

The function A is another example of a transit function that can be characterized by transit axioms only. This was the main result in [3].

Theorem 7. *Let $R : V \times V \to 2^V$ be a transit function satisfying the extra axioms*

$$(a4) \quad w \in R(u, v) \Rightarrow R(w, v) \subseteq R(u, v);$$

$$(a5) \quad R(u, x) \cap R(x, v) = \{x\} \Rightarrow R(u, x) \cup R(x, v) = R(u, v);$$

$$(a6) \quad R(u, v) \not\subseteq R(u, w) \Rightarrow \exists x \in R(u, v) \text{ with } x \neq u \text{ such that } R(u, x) \cap R(x, w) = \{x\}.$$

Then R is the all-paths transit function of the transit graph of R.

4.5 *The cut-vertex transit function*

Another way to analyze the block structure of a graph is to study the cut-vertices. For a connected graph $G = (V, E)$ we define the transit function C by

$$C(u, u) = \{u\},$$

$$C(u, v) = \{u, v\} \cup \{w | w \text{ is a cut-vertex between } u \text{ and } v\}.$$

Note that $C(u, v) = F_A(u, v)$, where A is the above all-paths function of G.

Clearly, $C(u, v) = \{u, v\}$ if and only if u and v are in the same block. The C-convexity is as follows. A set W is C-convex if and only if every cut-vertex between any two elements of W also lies in W. This is another example of a transit function R such that G and G_R do not coincide. We list some obvious first results on C.

Proposition 8. *Let G be a connected graph. Then C is a betweenness on G.*

Proposition 9. *Let G be a connected graph. Then G_C is the block closure of G.*

Proposition 10. *Every connected graph is C-monotone.*

Proposition 11. *Let G be a connected graph. Then $|C(u, v, w)| \leq 1$, for any three vertices u, v, w of G.*

Proof. If at least two of u, v, w lie in the same block, say u and v, then $C(u, v, w) \subseteq \{u, v\}$. Clearly, u and v can not simultaneously be cut-vertices between v, w and u, w respectively. Hence $C(u, v, w)$ is either empty or a singleton. So suppose that all three of u, v, w are in different blocks. First, one of the three, say v, might be in a block that is "in between" u and w. Now $C(u, v) \cap C(v, w) = \{v\}$, and we are done. Otherwise, there is either a unique cut-vertex x or a unique block B "in between" any pair of the three. In the latter case it is easy to see that the intersection $C(u, v, w)$ is empty. □

Proposition 12. *Let G be a connected graph. Then $|C(u, v, w)| = 1$ for any triple of vertices of G if and only if G is a tree.*

Proof. If G is a tree, then $C(u, v) = I(u, v)$, and we are done. If G is not a tree, then it contains a block with at least three distinct vertices, say u, v, w. Then we have $C(u, v, w) = \emptyset$. □

The transit function C is an instance of a transit function with $F_C = R_C = C$.

4.6 *The prefiber transit function*

For u and v in G, the prefiber interval between u and v is the set

$$F(u, v) = \{w | I(u, w) \cap I(w, v) = \{w\}\}.$$

This notion of interval was introduced by Nebeský [24] to characterize the block graphs, in his words Husimi trees. Block graphs are special instances of quasi-median graphs. These graphs are isometric subgraphs of Hamming graphs. As observed above, F plays a role in one of the characterizations of Hamming graphs, see [19]. Many of the prototype problems and questions turn out to be quite interesting for the prefiber transit function, see [21].

4.7 *The longest path transit function*

Let G be a connected graph. The *longest-path transit function* L on G is the transit function defined by

$$L(u, v) = \{w | w \text{ lies on a longest path between } u \text{ and } v\}.$$

Clearly, $L(u, u) = \{u\}$, for any u. One may expect that this path transit function will have some very different and unexpected properties. Take for example, the graph consisting of two vertices u and v, three internally disjoint paths of length four between u and v, and a vertex x adjacent to u and v only. Then the L-convex sets are the trivial sets (i.e. the empty set, the singletons, and the whole vertex set) and the set of all vertices except x.

Let $\lambda(u, v)$ be the length of a longest u, v-path. Then it is easy to see that λ defines a metric on V. What properties does this metric have?

4.8 *Single paths transit functions*

Let G be a connected graph. Let Φ be a set of paths in G such that, for any two vertices u and v in G, there exits a unique path $P_{u,v}$ between u and v in Φ. We call such a path family a *single-path family*, and its Φ-path transit function R a single-path transit function.

If G is a geodetic graph, then, by definition, its interval function I is a single-path transit function. A characterization of geodetic graphs involving the interval function is given by Nebeský [27].

Another example is the following. Let G be a cycle, and let uv be a fixed edge of G. Let $P_{u,v}$ be the u, v-path containing the single edge uv. For any pair of vertices x, y distinct from u, v, let $P_{x,y}$ be the x, y-path in G not containing the edge uv. Then this set of paths is a single-path family, and its transit function R is a single-path transit function. Note that this transit function has some nice properties: the cycle is R-monotone, and R is a betweenness.

In addition to the prototype problems and questions that may be relevant, we mention just one other question.

Open Problem 4.7.1. Characterize the graphs for which the longest path transit function is a single-path transit function.

5. Transit functions on posets

Many results on the interval function I have their analogue in the area of partially ordered sets, semilattices and lattices. A prime example of this is the theory of median graphs. Let $G = (V, E)$ be a median graph, and let u be a vertex of G. If we define the ordering $\leq_u$ on V by $v \leq_u w$ if $v \in I(u, w)$, then the partially ordered set $(V, \leq_u)$ is a so-called median semilattice, see [19]. On the other hand, the Hasse graph of a median semilattice is a median graph. Here the Hasse graph of a poset is the Hasse diagram where the orientation is ignored. This relationship makes it possible to translate many results on the interval function of median graphs to their analogues of median semilattices. There are generalizations possible to partially ordered sets in general. It leads much too far to give a detailed presentation here. But it becomes quite clear that many of the above problems and ideas have their counterpart in the area of partially ordered sets. Not all of these are just a trivial transfer of a result on graphs into a result on posets. So, again, the transit function approach may provide us with many new ideas and problems to be pursued for their own sake. An early instance of relevant results, even on other structures than posets, is the work of Sholander, see [31,32]. A first and interesting step in this area is represented by [15].

References

[1] S. P. Avann, Metric ternary distributive semi-lattices, *Proc. Amer. Math. Soc.*, **12** (1961) 407–414.

[2] H.-J. Bandelt, H. M. Mulder and E. Wilkeit, Quasi-median graphs and algebras, *J. Graph Theory*, **18** (1994) 681–703.

[3] M. Changat, S. Klavžar and H. M. Mulder, The all-paths transit function of a graph, *Czechoslovak J. Math.*, **51** (2001) no. 126, 439–447.

[4] M. Changat and J. Mathews, On triangle path convexity in graphs, *Discrete Math.*, **206** (1999) 91–95.

[5] M. Changat and J. Mathews, Interval monotone graphs: minimal path convexity, in: Proceedings of the Conference on Graph Connections, (eds.) R. Balakrishnan, H. M. Mulder and A. Vijayakumar, Allied Publishers Ltd, New Delhi (1999) 87–90.

[6] M. Changat and J. Mathews, Induced path transit function, monotone and Peano axioms, *Discrete Math.*, **286** (2004) 185–194.

[7] M. Changat and J. Mathews, Characterizations of J-monotone graphs, (in this volume) 47–55.

[8] M. Changat, J. Mathews and H. M. Mulder, Induced path transit function, betweenness and monotonicity, *Electronic Notes in discrete math.*, **15** (2003) 62–65.

[9] M. Changat, J. Mathews and H. M. Mulder, The induced path function, monotonicity and betweennes, Report EI 2006–23, Econometrisch Instituut, Erasmus Universiteit (2006) 14 pp.

[10] M. Changat, H. M. Mulder and G. Sierksma, Convexities related to path properties on graphs. *Discrete Math.*, **290** (2005) 117–131.

[11] P. Duchet, Convex sets in Graphs II. Minimal path convexity, *J. Combin. Theory Ser. B*, **44** (1988) 307–316.

[12] A. J. Goldman and C. J. Witzgall, A localisation theorem for optimal facility location, *Transportation Science*, **4** (1970) 406–409.
[13] M. S. Jacobson, F. R. McMorris and H. M. Mulder, An introduction to tolerance intersection graphs, in: (eds.) Y. Alavi e.a. (1990).
[14] S. Klavžar and H. M. Mulder, Median graphs: characterizations, location theory, and related structures, *J. Combin. Math. Combin. Comput.*, **30** (1999) 103–127.
[15] A. Mathews and J. Mathews, Transit functions on posets and lattices, (in this volume) 105–116.
[16] F. R. McMorris, H. M. Mulder and F. S. Roberts, The median procedure on median graphs, *Discrete Appl. Math.*, **84** (1998) 165–181.
[17] M. A. Morgana and H. M. Mulder, The induced path convexity, betweenness, and svelte graphs, *Discrete Math.*, **254** (2002) 349–379.
[18] H. M. Mulder, The structure of median graphs, *Discrete Math.*, **24** (1978) 197–204.
[19] H. M. Mulder, The interval function of a graph, MC-tract 132, Mathematisch Centrum, Amsterdam (1980).
[20] H. M. Mulder, Metric graph theory. The case of median graphs, in: Proceedings of the Conference on Graph Connections, (eds.) R. Balakrishnan, H. M. Mulder and A. Vijayakumar, Allied Publishers Ltd, New Delhi (1999) 19–32.
[21] H. M. Mulder, The prefiber convexity and its transit function, in preparation.
[22] H. M. Mulder and A. Schrijver, Median graphs and Helly hypergraphs, *Discrete Math.*, **25** (1979) 41–50.
[23] L. Nebeský, Median graphs, *Comment. Math. Univ. Carolinae*, **12** (1971) 317–325.
[24] L. Nebeský, Algebraic properties of Husimi trees, *Časopis Pěst. Mat.*, **107** (1982) 116–123.
[25] L. Nebeský, A characterization of the interval function of a connected graph, *Czechoslovak Math. J.*, **44** (1994) 173–178.
[26] L. Nebeský, Characterizing the interval function of a connected graph, *Mathematica Bohemica*, **123** (1994) 137–144.
[27] L. Nebeský, The interval function of a connected graph and a characterization of geodetic graphs, *Math. Bohemica*, **126** (2001) 247–254.
[28] L. Nebeský, The induced paths in a connected graph and a ternary relation determined by them, *Math. Bohemica*, **127** (2002) 397–408.
[29] P. G. Narasimha-Shenoi, On a metric related transit function and its betweenness (in this volume) 73–79.
[30] E. Sampathkumar, B-systems, in: Graph Theory and its Applications, Proceedings of the National Workshop, (eds.) S. Arumugam, B. D. Acharya and E. Sampathkumar, Manonmaniam Sundaranar University, Tirunelveli (1996) 173–185.
[31] M. Sholander, Trees, lattices, order, and betweenness, *Proc. Amer. Math. Soc.*, **3** (1952) 369–381.
[32] M. Sholander, Medians, lattices, and trees, *Proc. Amer. Math. Soc.*, **5** (1954) 808–812.
[33] C. Tardif, Prefibers and the Cartesian product of metric spaces, *Discrete Math.*, **109** (1992) 283–288.
[34] M. van de Vel, Theory of convex structures, North Holland, Amsterdam (1993).

Joint Proc. Int. Wrk.shps. Conv. in Discr. Struct. and Metric & Conv. Graph Th. No. 5, 2008, pp. 131–142.

Generalizing the Krein-Milman Property in Graph Convexity Spaces: A Short Survey

Ignacio M. Pelayo

Universitat Politècnica de Catalunya, Barcelona, Spain
e-mail: ignacio.m.pelayo@upc.edu

Abstract. A *convexity* on a non-empty finite set V is a family $\mathcal{C}$ of subsets of V (to be regarded as convex sets), which is closed under intersections and which contains both V and the empty set. The pair $(V, \mathcal{C})$ is called a *convexity space*. The smallest convex set containing a set $A \subseteq V$ is said to be the *convex hull* of A. Given a convex set $W \subseteq V$, a vertex $v \in W$ is called an *extreme vertex* of W if the set $W \backslash \{v\}$ is also convex. A *convex geometry* is a convexity space satisfying the so-called *Krein-Milman* property: Every convex set of V is the convex hull of its extreme vertices. A *Krein-Milman type property* is any property similar to the previous one in which either the set of vertices (the extreme vertex set) or the operator (the convex hull one), or both, are replaced. In this paper, we survey the state-of-art of these type of properties for graph convexity spaces.

Keywords. Boundary, contour, extreme vertex, geodetic set, graph convexity space, hull set, Krein-Milman property, monophonic set, Steiner set.

1. Introduction

All graphs in this paper are finite, undirected, simple and connected. For undefined basic concepts we refer the reader to [32].

A *convexity* on a non-empty set V is a family $\mathcal{C}$ of subsets of V (to be regarded as convex sets), such that:

(C1) $\emptyset, V \in \mathcal{C}$.

(C2) Arbitrary intersections of convex sets are convex.

(C3) Every nested union of convex sets is convex.

The pair $(V, \mathcal{C})$ is called a *convexity space* (see [31]). The smallest convex set containing a set $A \subseteq V$ is denoted by $[A]_{\mathcal{C}}$ and is called the *convex hull* of A. A *graph convexity space* is an ordered pair $(G, \mathcal{C})$ formed from a connected graph $G = (V, E)$, and a convexity $\mathcal{C}$ on V such that $(V, \mathcal{C})$ is a convexity space satisfying the following additional axiom:

(C4) Every member of $\mathcal{C}$ induces a connected subgraph of G.

Let V be a set and $I : V \times V \to 2^V$ be a mapping such that

(I1) $x, y \in I(x, y)$ for every $x, y \in V$.

(I2) $I(x, y) = I(y, x)$ for every $x, y \in V$.

The *I-closed* subsets of V are those subsets $C \subseteq V$ such that $I(C \times C) = C$. The set $\mathcal{C}_I$ of I-closed subsets satisfies axioms (C1), (C2) and (C3) of convexity spaces. The function I is called an *interval-function* of the convexity space $(V, \mathcal{C}_I)$. A convexity space $(V, \mathcal{C})$ admitting an interval-function I such that $\mathcal{C} = \mathcal{C}_I$ is named an *interval convexity space* $(V, I, \mathcal{C}_I)$ (see [5]).

The most natural convexities in a graph are path convexities (a type of interval convexity) defined by a system $\mathcal{P}$ of paths (see [8]). For vertices u and v, the $\mathcal{P}$-interval $\mathcal{P}(u, v)$ is the set of all vertices w belonging to paths from $\mathcal{P}$ connecting u and v. The so-called *coarsest convexity* is obtained when $\mathcal{P}$ is the set of all paths (see [6]). The *triangle-path convexity* is obtained when $\mathcal{P}$ is the set of all triangle paths (i.e., paths allowing only short chords) [7]. The *total convexity* is that which is obtained when $\mathcal{P}$ is the set of all triangle-free paths (i.e., paths allowing just long chords) [13]. The *monophonic convexity* is that which is obtained when $\mathcal{P}$ is the set of all induced paths [15,18]. The m^3*-convexity*[1] is that which is obtained when $\mathcal{P}$ is the set of all induced paths of length at least 3 [14]. The *longest path convexity* is obtained when $\mathcal{P}$ is the set of all longest paths [11]. Finally, the *geodesic convexity*, also called *metric convexity*, is obtained when $\mathcal{P}$ is the set of all shortest paths [18,19,30].

Given a graph convexity space $(G = (V, E), \mathcal{C})$ and a convex set $W \subseteq V$, a vertex $v \in W$ is called an *extreme vertex* of W if the set $W \backslash \{v\}$ is also convex. The graph G is called a *convex geometry* (or an *antimatroid*) if it satisfies the *Krein-Milman* property: *Every convex set of G is the convex hull of its extreme vertices* [18,23,24]. Certainly, this condition can be seen as a procedure that may allow us to recover any convex set from its extreme vertices, by means of the convex hull operator. From this point of view, the interest of any similar property is that a small subset of any convex set retains all the information of the whole set.

A vertex v of a connected graph G is a *simplicial vertex* if the subgraph induced by its neighborhood is a clique, i.e., if v is not the midpoint of an induced path P_3. A vertex v of G is a *semisimplicial vertex* if it is not a midpoint of an induced path P_4. Notice that for both the geodesic and the monophonic convexity, a vertex is simplicial if and only if it is an extreme vertex of V (see [17]). Mainly for this reason, the set of all simplicial vertices of a connected graph G is usually called the *extreme set* of G and denoted by $Ext(G)$. Observe also that for the m^3-convexity, a vertex is semisimplicial if and only if it is an extreme vertex of V.

A graph G is *chordal* if it contains no induced cycle of length greater than 3. A chordal graph is called *Ptolemaic* if it is distance-hereditary, that is, if every chordless path is a shortest path. Farber and Jamison [18] characterized both chordal and Ptolemaic families, in terms of the Krein-Milman property[2].

Theorem 1 ([18]). *The geodesic convexity of a graph G is a convex geometry if and only if G is Ptolemaic.*

Theorem 2 ([18]). *The monophonic convexity of a graph G is a convex geometry if and only if G is chordal.*

[1]Note that this convexity does not satisfy axiom (C4), i.e., a m^3-convex set is not necessarily connected.

[2]In this paper, and also in [14], this property is called the *Minkowski-Krein-Milman property*.

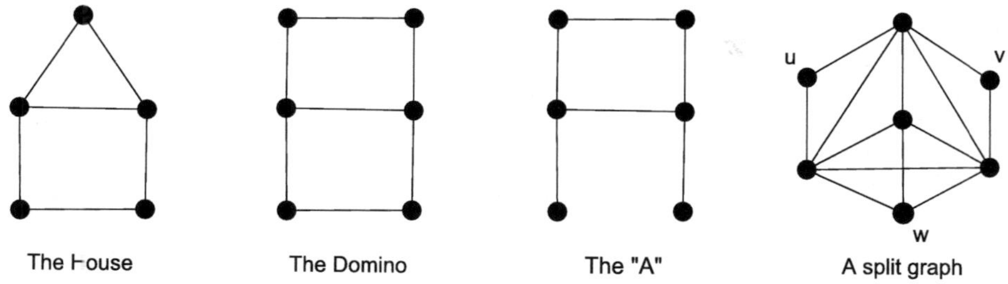

Figure 1. Some basic graphs.

An *HHD-free graph* is a graph which contains no hole (i.e., no cycle of order greater than 4), house or domino as an induced subgraph. A *weak bipolarizable graph* is an HHD-free graph which contains no "A" as an induced subgraph (see Figure 1). In [14], the weak bipolarizable family was characterized in terms of the Krein-Milman property.

Theorem 3 ([14]). *The m^3-convexity of a graph G is a convex geometry if and only if G is weak bipolarizable.*

Observe that every induced subgraph of a chordal (resp. Ptolemaic) (resp. weak bipolarizable) graph is chordal (resp. Ptolemaic) (resp. weak bipolarizable). This means that the necessity of the above theorems is simply a corollary of the following ones.

Theorem 4. *If $G = (V, E)$ is a Ptolemaic graph, then V is the convex hull of the set of its extreme vertices, when considering the geodesic convexity.*

Theorem 5. *If $G = (V, E)$ is a chordal graph, then V is the convex hull of the set of its extreme vertices, when considering the monophonic convexity.*

Theorem 6. *If $G = (V, E)$ is a weak bipolarizable graph, then V is the convex hull of the set of its extreme vertices, when considering the m^3-convexity.*

Thus, we could think of extending these properties in two different ways. In the one instance, recovering convex sets on larger graph classes and, in the other instance, using an operator simpler than the convex hull one, such as for example, the geodetic closure operator I (or the monophonic closure operator J). In both cases, determining the defining properties of sets of vertices that play a similar role to that of extreme vertices is necessary. The rest of this paper is devoted to surveying the work done concerning these kind of properties, known as *Krein-Milman type properties*, within the ambit of both the geodesic and monophonic convexities.

2. Geodesic and monophonic convexities

2.1 *Geodesic convexity*

Given vertices u, v in a graph $G = (V, E)$, the *geodetic interval* $I[u, v]$ is the set of all vertices belonging to a shortest $u - v$ path. Given $W \subseteq V$, the *geodetic closure*

$I[W]$ of W is the union of all geodetic intervals $I[u, v]$ taken over all pairs $u, v \in W$, i.e., $I[W] = \cup_{u,v\in W} I[u, v]$. Similarly, the *edge geodetic interval* $I_e[u, v]$ is the set of all edges belonging to a shortest $u - v$ path. Given $W \subseteq V$, the *edge geodetic closure* $I_e[W]$ of W is the union of all edge geodetic intervals $I_e[u, v]$ taken over all pairs $u, v \in W$, i.e., $I_e[W] = \cup_{u,v\in W} I_e[u, v]$.

A set of vertices $W \subseteq V$ is called *geodesically convex* (or simply *g-convex*) if $W = I[W]$. The smallest g-convex set containing W is denoted by $[W]_g$ and is called the *g-convex hull* of W. Certainly, $W \subseteq I[W] \subseteq [W]_g \subseteq V$. A set of vertices $W \subseteq V$ is called a *g-hull set* if $[W]_g = V$, a *geodetic set* if $I[W] = V$, and an *edge geodetic set* if $I_e[W] = E$. Clearly, every nonempty edge geodetic set is geodetic and every geodetic set is a g-hull set.

A g-hull set of G of minimum cardinality is a minimum g-hull set and its cardinality is the *g-hull number* $hn(G)$ [10,17]. Similarly, the *geodetic number* $gn(G)$ is the minimum cardinality of a geodetic set. Certainly, $hn(G) \leq gn(G)$.

2.2 *Monophonic convexity*

Given vertices u, v in a graph $G = (V, E)$, a $u - v$ chordless path is an induced path joining u and v. The *monophonic interval* $J[u, v]$ is the set of all vertices belonging to some chordless $u - v$ path. Given $W \subseteq V$, the *monophonic closure* $J[W]$ of W is the union of all monophonic intervals $J[u, v]$ taken over all pairs $u, v \in W$, i.e., $J[W] = \cup_{u,v\in W} J[u, v]$. Similarly, the *edge monophonic interval* $J_e[u, v]$ is the set of all edges belonging to some chordless $u - v$ path. Given $W \subseteq V$, the *edge monophonic closure* $J_e[W]$ of W is the union of all edge monophonic intervals $J_e[u, v]$ taken over all pairs $u, v \in W$, i.e., $J_e[W] = \cup_{u,v\in W} J_e[u, v]$.

A set of vertices $W \subseteq V$ is called *monophonically convex* (see [18]), or simply *m-convex*, if $W = J[W]$. The smallest m-convex set containing W is denoted by $[W]_m$ and is called the *m-convex hull* of W. Certainly, $W \subseteq I[W] \subseteq J[W] \subseteq [W]_m \subseteq V$. A set of vertices $W \subseteq V$ is called an *m-hull set* if $[W]_m = V$, a *monophonic set* if $J[W] = V$, and an *edge monophonic set* if $J_e[W] = E$. Notice that, every edge monophonic set is monophonic and every monophonic set is an m-hull set. Observe also that, every edge geodetic set is edge monophonic, every geodetic set is monophonic, and every g-hull set is an m-hull set.

An m-hull set of G of minimum cardinality is a minimum m-hull set and its cardinality is the m-hull number $mhn(G)$. The monophonic number $mn(G)$ is the minimum cardinality of a monophonic set. Certainly, $mhn(G) \leq mn(G) \leq gn(G)$. It is also easy to see that neither $mn(G) \leq hn(G)$ nor $hn(G) \leq mn(G)$ is always true (see Table 1). Going one step further, the following realization theorem was proved.

Theorem 7 ([21]). *For any integers a, b, c, d such that $3 \leq a \leq b \leq c \leq d$, there exist connected graphs G_1 and G_2 such that:*

1. $a = mhn(G_1)$, $b = mn(G_1)$, $c = hn(G_1)$, *and* $d = gn(G_1)$,
2. $a = mhn(G_2)$, $b = hn(G_2)$, $c = mn(G_2)$, *and* $d = gn(G_2)$.

Table 1. Convexity parameters of some basic families of graphs.

G	P_n	C_{2l}	C_{2l+1}	T_n	K_n	$K_{p,q}$ $(2 \le p \le q)$	$W_{1,p}$ $(p \ge 4)$	Q_n
$mhn(G)$	2	2	2	♯ leaves	n	2	2	2
$mn(G)$	2	2	2	♯ leaves	n	$\min\{4, p\}$	2	2
$hn(G)$	2	2	3	♯ leaves	n	2	$\lceil \frac{p}{2} \rceil$	2
$gn(G)$	2	2	3	♯ leaves	n	$\min\{4, p\}$	$\lceil \frac{p}{2} \rceil$	2

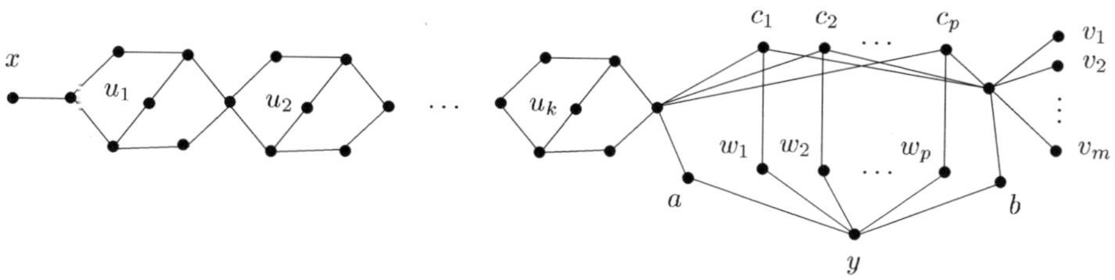

Figure 2. In this graph, $3 \le hn(G) \le st(G) < gn(G)$, since $W_1 = \{x, y\} \cup \{v_i\}_{i=1}^m$ is a minimum g-hull set, $W_2 = W_1 \cup \{u_i\}_{i=1}^k$ is a minimum Steiner set and $W_3 = W_2 \cup \{w_i\}_{i=1}^p$ is a minimum geodetic set.

3. Extending the extreme set

3.1 *Steiner sets*

For a nonempty subset W of the vertices of a connected graph G, a connected subgraph of G with the minimum number of edges that contains all of W clearly must be a tree; such a tree is called a *Steiner W-tree.* The *Steiner distance* $d_S(W)$ of W is the size of a Steiner W-tree. The *Steiner interval* $S(W)$ of W consists of all vertices that lie on some Steiner W-tree. If $S(W) = V(G)$, then W is called a *Steiner set* for G. Notice that, for every Steiner set W, $Ext(G) \subseteq W$.

The *Steiner (geodetic) number*[3] $st(G)$ of G is defined as the minimum cardinality of a Steiner set of G [12]. The *Steiner set problem* for G is concerned with Steiner sets in G and the Steiner number $st(G)$ of G.

A set $W \subseteq V(G)$ is an *edge Steiner set* if the edges lying in some Steiner W-tree cover $E(G)$. Notice that every edge Steiner set is a Steiner set. It is easy to find examples where the converse of this statement is not true.

As was shown in [29], a Steiner set need not be a geodetic or a g-hull set (see Figure 2). Going one step further, the following realization theorem was proved.

Theorem 8 ([20]). *For every triple a, b, c of integers with $3 \le a \le b \le c$, there exist connected graphs G_1, G_2 and G_3 such that:*

1. $hn(G_1) = a,\ gn(G_1) = b,\ st(G_1) = c$
2. $hn(G_2) = a,\ st(G_2) = b,\ gn(G_2) = c$
3. $st(G_3) = a,\ hn(G_3) = b,\ gn(G_3) = c$

[3] In [27], the term *Steiner number* was previously introduced by only considering those sets W such that $d_S(W) = |V| - 1$.

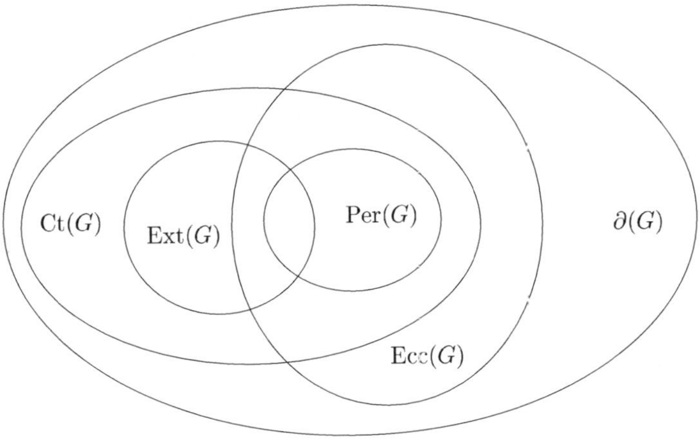

Figure 3. Boundary-type sets.

3.2 *Boundary vertices*

Let $G = (V, E)$ be a connected graph and $u, v \in V$. The vertex v is said to be a *boundary vertex* of u if no neighbor of v is further away from u than v [9]. By $\partial(u)$ we denote the set of all boundary vertices of u. A vertex v is called a *boundary vertex of G* if $v \in \partial(u)$ for some vertex $u \in V(G)$. The *boundary* $\partial(G)$ of G is the subgraph induced by all boundary vertices of G.

Given $u, v \in V$, the vertex v is called an *eccentric vertex of* u if no vertex in V is further away from u than v, that is, if $d(u, v) = ecc(u) = \max\{d(u, v)|v \in V\}$. A vertex v is called an *eccentric vertex* of G if it is the eccentric vertex of some vertex $u \in V$. The *eccentric subgraph* $Ecc(G)$ of G is the subgraph induced by all eccentric vertices of G [9].

A vertex $v \in V$ is called a *contour vertex* of G if no neighbor of v has eccentricity greater than $ecc(v)$ [4]. The contour $Ct(G)$ of G is the subgraph induced by all contour vertices of G.

A vertex $v \in V$ is called a *peripheral vertex* of G if no vertex in V has eccentricity greater than $ecc(v)$, that is, if the eccentricity of v is exactly equal to the diameter of G. The *periphery* $Per(G)$ of G is the subgraph induced by all peripheral vertices of G.

For simplicity, we also use $\partial(G)$, $Ecc(G)$, $Ct(G)$ and $Per(G)$ to denote the respective collections of vertices. Notice that every simplicial vertex is a contour vertex, i.e., $Ext(G) \subseteq Ct(G)$. It is also clear that: $Per(G) \subseteq Ct(G) \cap Ecc(G)$ and $Ecc(G) \cup Ct(G) \subseteq \partial(G)$ (see Figure 3).

3.2.1 *On the sizes of sets of boundary vertices*

The cardinalities of the boundary vertex sets of any graph G must satisfy certain nontrivial constraints, specified by Theorem 9, proved in [3].

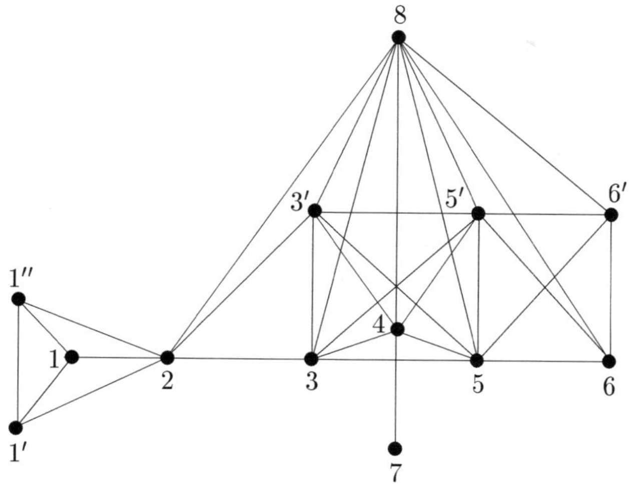

Figure 4. In this graph G: $Per(G) = \{1, 1', 1'', 7\}$, $Ct(G) = \{1, 1', 1'', 5, 5', 6, 6', 7\}$, $Ext(G) = Ecc(G) = \{1, 1', 1'', 6, 6', 7\}$, and $\partial(G) = \{1, 1', 1'', 3, 3', 5, 5', 6, 6', 7\}$.

Theorem 9 ([3]). *Let $G = (V, E)$ be a connected graph.*

1. *If $|Per(G)| = |Ct(G)| = 2$, then either $|\partial(G)| = 2$ or $|\partial(G)| \geq 4$.*
2. *If $|Ecc(G)| = |Per(G)| + 1$, then $|\partial(G)| > |Ecc(G)|$.*

As a direct consequence of these properties, the following result is immediately derived.

Corollary 1 ([3]). *Let G be a nontrivial connected graph such that $|Per(G)| = a$, $|Ct(G)| = b$, $|Ecc(G)| = c$ and $|\partial(G)| = d$. Then,*

$$\begin{cases} 2 \leq a \leq \min\{b, c\} \leq \max\{b, c\} \leq d, \\ (a, b, c, d) \neq (2, 2, 2, 3), \\ (a, b, c, d) \neq (a, b, a + 1, a + 1). \end{cases} \tag{1}$$

Starting from this result, the following realization theorems have been proved.

Theorem 10 ([3]). *Let $(a, b, c, d) \in \mathbb{Z}^4$ be integers satisfying the constraints (1) of Corollary 1. Then, there exists a connected graph $G = (V, E)$ such that $|Per(G)| = a$, $|Ct(G)| = b$, $|Ecc(G)| = c$, and $|\partial(G)| = d$.*

Theorem 11 ([2]). *Let $(a, b, c, d) \in \mathbb{Z}^4$ be integers satisfying the constraints (1) of Corollary 1. Then, there exists a connected chordal graph $G = (V, E)$ such that $|Per(G)| = a$, $|Ct(G)| = b$, $|Ecc(G)| = c$, $|\partial(G)| = d$.*

Theorem 12 ([2]). *Let $(a, b, c, d) \in \mathbb{Z}^4$ be integers satisfying the constraints (1) of Corollary 1. Then, there exists a connected chordal graph G such that $|Per(G)| = a$, $|Ext(G)| = b$, $|Ecc(G)| = c$, $|\partial(G)| = d$.*

Lemma 1 ([22]). *Let G be a connected graph. Then,*

1. $|\partial(G)| = 2$ *if and only if G is a path.*
2. *If* $|\partial(G)| = 3$, *then* $\partial(G) = Ct(G) = Ext(G)$.

From this last result, the following corollary is immediately derived.

Corollary 2 ([22]). *Let G be a connected graph with* $|Ext(G)| = a$, $|Per(G)| = b$, $|Ct(G)| = c$, *and* $|\partial(G)| = d$. *Then*

1. $0 \leq a \leq c \leq d$, *and* $2 \leq b \leq c \leq d$,
2. $(a, b, c, d) \neq (a, 2, 2, 3)$ *for any value of a,*
3. $(a, b, c, d) \neq (2, 2, 3, 3)$,
4. $(a, b, c, d) \neq (2, 3, 3, 3)$.

Finally, the following realization theorem has been proved.

Theorem 13 ([22]). *Let* $(a, b, c, d) \in \mathbb{Z}^4$ *be integers satisfying the constraints of Corollary 2. If* $a \geq 2$, *then there exists a connected graph G such that* $|Ext(G)| = a$, $|Per(G)| = b$, $|Ct(G)| = c$, *and* $|\partial(G)| = d$.

4. Krein-Milman type properties

4.1 *Steiner sets*

It has already been mentioned that not every Steiner set is geodetic. Even if we restrict ourselves to the family of chordal graphs, this statement is still true. For example, the subset of vertices $\{u, v, w\}$ of the split graph illustrated in Figure 1 is a Steiner set but it is not geodetic. On the other hand, within certain classes of graphs, every Steiner set is, in fact, geodetic.

Theorem 14 ([16,20,28]). *If a connected graph G is either distance hereditary, or 3-Steiner distance hereditary or an interval graph, then every Steiner set of G is geodetic.*

Theorem 15 ([21]). *In the class of interval graphs, every edge Steiner set is edge geodetic.*

Observe that all of these results can be rewritten in a similar way to the statements of Theorem 4 and 6. For example, Theorem 14 can be restated as follows: *If* $G = (V, E)$ *is a distance hereditary graph and W is a Steiner set of G, then V is the geodetic closure of W.*

The current state-of-art for the question: *For which chordal graphs is every Steiner set a geodetic set?* is displayed in Figure 5 (see [20]). The only Krein-Milman type properties for Steiner sets, valid for every graph, are to be found when considering the monophonic convexity.

Theorem 16 ([20]). *Every Steiner set is monophonic.*

Theorem 17 ([21]). *Every edge Steiner set is edge monophonic.*

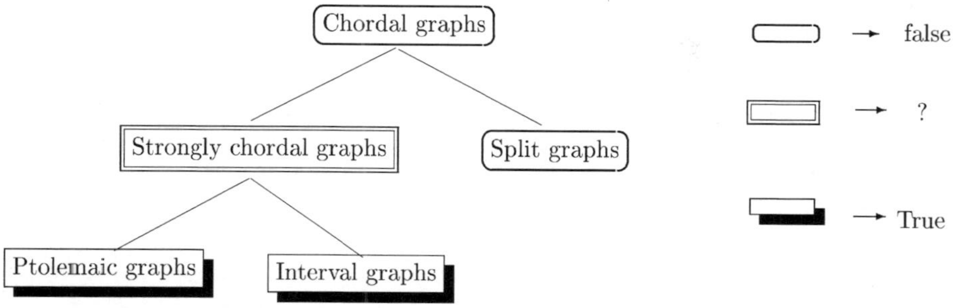

Figure 5. For which chordal graphs is every Steiner set a geodetic set?

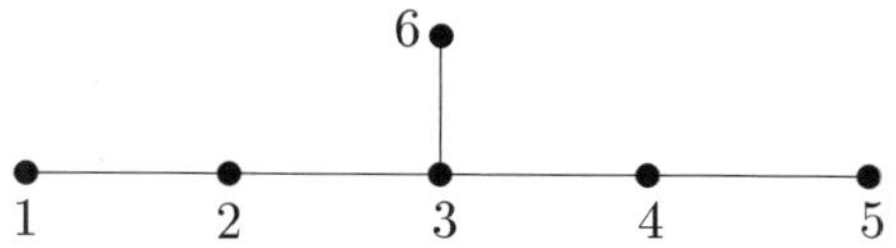

Figure 6. Tree T for which $Ecc(T) = \{1, 5\}$ and $[Ecc(T)]_g = V(T) - 6$.

4.2 *Boundary vertices*

A number of general statements that involve different sets of boundary vertex sets and their relation to geodetic and edge geodetic sets have been obtained.

Theorem 18 ([22]). *Let $G = (V(G), E(G))$ be a connected graph and $u \in V(G)$. Then $S = \{u\} \cup \partial(u)$ is a geodetic set. Moreover, if G is bipartite then $S = \{u\} \cup \partial(u)$ is also an edge geodetic set.*

Theorem 19 ([22]). *The boundary of every graph is edge geodetic.*

As was noticed in [4], the eccentric subgraph of a graph is in general far from being a g-hull set, even for the class of trees, as is shown in Figure 6.

As was shown in [4], the contour of a graph does not need to be geodetic, (see Figure 7) Nevertheless, it is always both a g-hull set and monophonic.

Theorem 20 ([4]). *The contour of every graph $G = (V, E)$ is a g-hull set.*

Theorem 21 ([21]). *The contour of every graph is monophonic.*

To finalize, we summarize a number of results involving the contour, that have been obtained by considering different kinds of constraints.

Theorem 22 ([3]). *If the contour of a graph is equal to its periphery, then it is geodetic.*

Theorem 23 ([3]). *Let G be a graph such that $Ct(G)\backslash Per(G) = \{y_1, \ldots, y_k\}$ and $ecc(y_i) = ecc(y_j)$, for each $i, j = 1, \ldots, k$. Then, $I[Ct(G)]$ is geodetic.*

Theorem 24 ([4,21]). *The contour of every distance-hereditary graph is geodetic.*

Theorem 25 ([2]). *The contour of every chordal graph is geodetic.*

Theorem 26 ([16]). *If G is either a 3-Steiner distance hereditary or a HHD-free graph, then its contour is geodetic.*

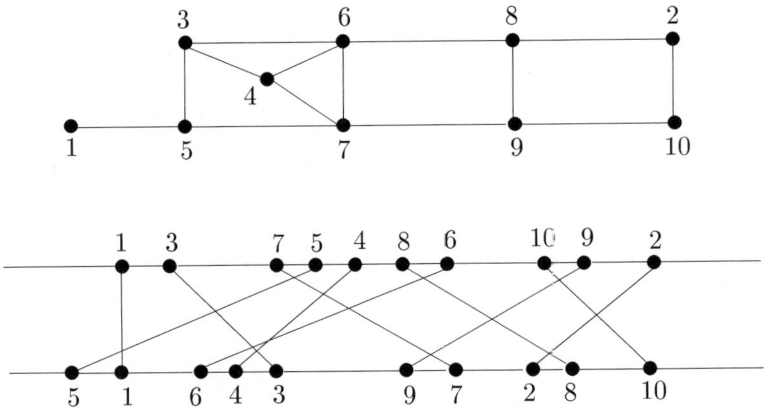

Figure 7. Permutation graph P of order 10 and its matching diagram (see [26]). Notice that $Ct(P) = \{1, 2, 3\}$ and $I[Ct(P)] = V(P) - 4$.

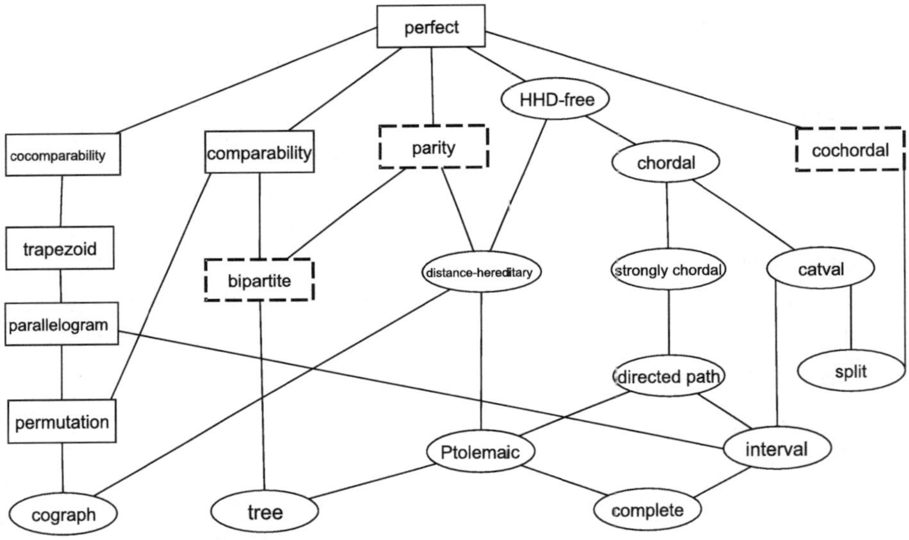

Figure 8. State-of-art of the question *For which perfect classes is the contour geodetic?*: Yes for solid ovals, not necessarily for solid squares, and open problem for dashed squares.

Acknowledgements

The author wishes to acknowledge the helpful comments and suggestions made by the anonymous referees.

References

[1] A. Brandstädt, Van B. Le and J. P. Spinrad, Graph classes: a survey, SIAM Monographs on Discrete Mathematics and Applications (1999).

[2] J. Cáceres, C. Hernando, M. Mora, I. M. Pelayo, M. L. Puertas and C. Seara, Geodeticity of the contour of chordal graphs, *Discrete Appl. Math.*, accepted.

[3] J. Cáceres, C. Hernando, M. Mora, I. M. Pelayo, M. L. Puertas and C. Seara, On geodetic sets formed by boundary vertices, *Discrete Math.*, **306** (2006) no. 2, 188–198.

[4] J. Cáceres, A. Márquez, O. R. Oellermann and M. L. Puertas, Rebuilding convex sets in graphs, *Discrete Math.*, **297** (2005) no. 1–3, 26–37.

[5] J. R. Calder, Some elementary properties of interval convexities, *J. London Math. Soc. (2)*, **3** (1971) 422–428.

[6] M. Changat, S. Klavžar and H. M. Mulder, The all-paths transit function of a graph, *Czechoslovak Math. J.*, **51(126)** (2001) no. 2, 439–448.

[7] M. Changat and J. Mathew, On triangle path convexity in graphs, *Discrete Math.*, **206** (1999) no. 1–3, 91–95.

[8] M. Changat, H. M. Mulder and G. Sierksma, Convexities related to path properties on graphs, *Discrete Math.*, **290** (2005) no. 2—3, 117–131.

[9] G. Chartrand, D. Erwin, G. L. Johns and P. Zhang, Boundary vertices in graphs, *Discrete Math.*, **263** (2003) no. 1–3, 25–34.

[10] G. Chartrand, F. Harary and P. Zhang, On the hull number of a graph, *Ars Combinatoria*, **57** (2000) 129–138.

[11] G. Chartrand, G. Johns and S. Tian, Detour distance in graphs. Quo vadis, graph theory?, 127–136, Ann. Discrete Math. 55, North-Holland, Amsterdam (1993).

[12] G. Chartrand and P. Zhang, The Steiner number of a graph, *Discrete Math.*, **242** (2002) no. 1–3, 41–54.

[13] V. Chepoi, Peakless functions on graphs, *Discrete Math.*, **73** (1997) no. 2, 175–189.

[14] F. F. Dragan, F. Nicolai and A. Brandstädt, Convexity and HHD-free graphs, *SIAM J. Discrete Math.*, **12** (1999) no. 1, 119–135.

[15] P. Duchet, Convex sets in graphs. II. Minimal path convexity, *J. Combin. Theory Ser. B*, **44** (1988) 307–316.

[16] L. Eroh and O. R. Oellermann, Geodetic sets in 3-Steiner distance hereditary graphs, preprint.

[17] M. G. Everett and S. B. Seidman, The hull number of a graph, *Discrete Math.*, **57** (1985) no. 3, 217–223.

[18] M. Farber and R. E. Jamison, Convexity in graphs and hypergraphs, *SIAM J. Algebraic Discrete Methods*, **7** (1986) no. 3, 433–444.

[19] F. Harary, E. Loukakis and C. Tsouros, The geodetic number of a graph, *Math. Comput. Modelling*, **17** (1993) no. 11, 89–95.

[20] C. Hernando, T. Jiang, M. Mora, I. M. Pelayo and C. Seara, On the Steiner, geodetic and hull numbers of graphs, *Discrete Math.*, **293** (2005) no. 1–3, 139–154.

[21] C. Hernando, M. Mora, I. M. Pelayo and C. Seara, On monophonic sets in graphs, preprint.

[22] C. Hernando, M. Mora, I. M. Pelayo and C. Seara, Some structural, metric and convex properties on the boundary of a graph, *Discrete Math.*, submitted.

[23] R. E. Jamison, Copoints in antimatroids, *Congr. Numer.*, **29** (1980) 535–544.

[24] B. Korte and L. Lovász, Homomorphisms and Ramsey properties of antimatroids, *Discrete Appl. Math.*, **15** (1986) no. 2–3, 283–290.

[25] T. Jiang, I. M. Pelayo and D. Pritikin, Geodesic Convexity and Cartesian Products in Graphs, *Networks*, accepted.

[26] T. A. McKee and F. R. McMorris, Topics in intersection graph theory, *SIAM Monographs on Discrete Mathematics and Applications* (1999).

[27] O. R. Oellermann, Steiner numbers in graphs, *Quaestiones Math.*, **13** (1990) no. 2, 159–164.

[28] O. R. Oellermann and M. L. Puertas, Steiner intervals and Steiner geodetic numbers in distance-hereditary graphs, *Discrete Math.*, **307** (2007) no. 1, 88–96.

[29] I. M. Pelayo, Comment on "The Steiner number of a graph" by G. Chartrand and P. Zhang, *Discrete Math.*, **242** (2002) 41–54, *Discrete Math.*, **280** (2004) no. 1–3, 259–263.

[30] V. P. Soltan, Metric convexity in graphs, *Studia Univ. Babeş-Bolyai Math.*, **36** (1991) no. 4, 3–43.
[31] M. Van de Vel, Theory of convex structures, North-Holland, Amsterdam (1993).
[32] D. B. West, Introduction to Graph Theory, Prentice-Hall, New Jersey (1999).

Author Index

Made in United States
Orlando, FL
04 November 2026

90206059R00085